Geometrical Vectors

Chicago Lectures in Physics Series
Robert M. Wald, series editor
Henry J. Frisch
Gene R. Mazenko
Sidney R. Nagel

Other *Chicago Lectures in Physics* titles available
from the University of Chicago Press

Currents and Mesons, by J. J. Sakurai (1969)
Mathematical Physics, by Robert Geroch (1984)
Useful Optics, by Walter T. Welford (1991)
Quantum Field Theory in Curved Spacetime and
  Black Hole Thermodynamics, by Robert M. Wald (1994)

Geometrical Vectors

Gabriel Weinreich

The University of Chicago Press

Chicago and London

The University of Chicago Press, Chicago 60637
The University of Chicago Press, Ltd., London
© 1998 by The University of Chicago
All rights reserved. Published 1998
Printed in the United States of America

16 15 14 13 12 11 10 09 08 07   5 6 7 8 9

ISBN-13: 978-0-226-89048-7 (paper)
ISBN-10: 0-226-89048-1 (paper)

LIBRARY OF CONGRESS CATALOGING-IN-PUBLICATION DATA

Weinreich, Gabriel.
    Geometrical vectors / Gabriel Weinreich.
        p.   cm. — (Chicago lectures in physics)
    Includes index.
    ISBN 0-226-89048-1 (pbk.)
    1. Vector analysis. 2. Geometry. I. Title. II. Series.
QC20.7.V4W45   1998
630.15'6182—dc21   97-51855
                                                    CIP

CONTENTS

PREFACE

Years of teaching Mathematical Methods of Physics at the University of Michigan to seniors and first-year graduate students convinced me that existing textbooks don't do an adequate job in the area of vector analysis: all too often, their treatment is a repetition of what students had already seen in earlier courses, with little or no insight into the essentially geometrical structure of the subject. For this reason, I got into the habit of substituting my own discussion for whatever the textbook contained, a few years ago even going so far as to distribute some informal notes under the title "Geometrical Vectors."

Reactions to those notes, on the part of both colleagues and students, were enthusiastic, and the thought of rewriting the material in a form suitable for publication naturally followed. It has, however, been clear to me from the beginning that the resulting book can only have an appreciable market if it is priced so that students can buy it *in addition* to one of the standard textbooks. Conversations with experts indicated that such a goal was, indeed, attainable in a paperback produced from my camera-ready copy, and so I got to work; three years later, the result of that work is before you.

In planning the book, I had to decide whether to call it "Geometrical Vectors" and limit its coverage correspondingly, or to go for "Geometrical Vectors and Tensors," discuss both subjects, and in doing so double its size. If in fact I opted for the first, it was not without a great deal of sadness, because tensor analysis is so beautiful, while at the same time its elementary treatments tend to be no better than those of vector analysis. Yet the practical demand for physicists to understand tensors is quite small compared to the ubiquitous use of vectors; and I was afraid that the consequent increase in price might make the resulting book inaccessible to what is, as I see it, its primary audience.

My enormous indebtedness in this work is to so many that it can only be acknowledged generically. And so my deep and heartfelt appreciation goes, first, to my students for constantly asking questions; second, to my teachers for teaching me to do the same; and last but by no means least, to my many colleagues (including a number of anonymous reviewers) in the crucible of whose conversation my present understanding was refined. I sincerely hope that others, too, may now find this understanding useful.

Gabriel Weinreich

Ann Arbor
December 1997

1

PROLOGUE:
WHAT THIS BOOK IS ABOUT

1.1 Introduction

We live in a three-dimensional flat space, or at least that is the way we usually think about it. It's true that special relativity motivates us to add time as a fourth dimension, and general relativity to consider our space as curved, that is, not subject to the constraint of having the angles of every triangle necessarily adding up to 180°. Nonetheless, the "normal," three-dimensional Euclidean space remains the space in which our intuition lives and breathes and has its being; it is, simply said, the only space that our mind can truly *picture*.

Because of this fact, it makes sense to concentrate our exploration of vectors in such a space, especially as there are some features of three-dimensional vectors − for example, the cross product − which have no exact equivalent in spaces of other dimensionality; we refer here to aspects that have nothing to do with any human ability to perceive or imagine, but arise out

1

of the mathematical nature of the space. (This does not mean, of course, that such features cannot be "generalized" to two, or four, or N dimensions, but such a generalization always carries a price, in that at least some of the properties of the concept in question have to be abandoned.) Nonetheless, the reader will find that, even though our treatment concentrates on the intuitively familiar three-dimensional flat Euclidean space, it will still provide a sufficiently solid grounding to begin generalizing to other types of spaces (see Chapter 9).

1.2 Where We Begin

In our discussions, we will repeatedly refer to the "traditional" treatment of vector analysis, meaning one in which vectors are always represented by arrows, and vector operations, although they may have a geometrical embodiment, are usually defined in terms of algebraic (or differential) operations on Cartesian components. Although it is our assumption that the reader has encountered such a treatment before, not only in a first-year physics course but perhaps even in one or more "intermediate" courses, we summarize its essential procedures and formulas before going any further.

Addition: Geometrically, two vectors A and B are added by the "parallelogram rule," according to which the tails of the two arrows are brought together, their parallelogram is completed, and a new vector C (which is the sum of A and B) is drawn from the common tail to the opposite vertex. Algebraically, the components of C are obtained by adding corresponding components of A and B:

$$\begin{aligned} C_x &= A_x + B_x \,, \\ C_y &= A_y + B_y \,, \\ C_z &= A_z + B_z \,. \end{aligned} \qquad (1.2.1)$$

Multiplication by a scalar: The length of the arrow is multiplied by the scalar, reversing direction if the scalar is negative. Algebraically, each component is multiplied by the scalar.

Scalar product of two vectors ("dot product"): The lengths of the two arrows are multiplied together, then further multiplied by the cosine of the angle between them. In terms of components,

$$\boldsymbol{A} \cdot \boldsymbol{B} = A_x B_x + A_y B_y + A_z B_z . \tag{1.2.2}$$

Vector product of two vectors ("cross product"): Given $\boldsymbol{A}$ and $\boldsymbol{B}$, a new vector is formed whose magnitude is the area of the parallelogram subtended by the two, whose direction is perpendicular to both $\boldsymbol{A}$ and $\boldsymbol{B}$, and whose sense is determined by a right-hand rule. In terms of components,

$$
\begin{aligned}
(\boldsymbol{A} \times \boldsymbol{B})_x &= A_y B_z - A_z B_y , \\
(\boldsymbol{A} \times \boldsymbol{B})_y &= A_z B_x - A_x B_z , \\
(\boldsymbol{A} \times \boldsymbol{B})_z &= A_x B_y - A_y B_x .
\end{aligned}
\tag{1.2.3}
$$

Differential operations on fields: Three operations of this category — gradient, divergence, and curl — are commonly defined in terms of components by the following formulas:

$$
\begin{aligned}
(\text{grad } \Phi)_x &= \partial\Phi/\partial x , \\
(\text{grad } \Phi)_y &= \partial\Phi/\partial y , \\
(\text{grad } \Phi)_z &= \partial\Phi/\partial z ;
\end{aligned}
\tag{1.2.4}
$$

$$\text{div } \boldsymbol{S} = \partial S_x/\partial x + \partial S_y/\partial y + \partial S_z/\partial z ; \tag{1.2.5}$$

$$
\begin{aligned}
(\text{curl } \boldsymbol{F})_x &= \partial F_z/\partial y - \partial F_y/\partial z , \\
(\text{curl } \boldsymbol{F})_y &= \partial F_x/\partial z - \partial F_z/\partial x , \\
(\text{curl } \boldsymbol{F})_z &= \partial F_y/\partial x - \partial F_x/\partial y .
\end{aligned}
\tag{1.2.6}
$$

Geometrical definitions are rarely, if ever, given.

By contrast, in this book we shall make it a point to stay away from algebraic operations as long as possible, not because there is anything wrong with them in principle but because they so thoroughly becloud the essentially geometrical nature of all vector notions which are, because of the way that human intuition is structured, much easier to understand. Indeed we shall try, at least at first, to formulate all definitions and operations not merely *geometrically* but *topologically,* that is, in terms which do not require numerical measurements of distances or angles. Another way of saying the same thing is that we shall search for relations that continue to maintain their validity even when space is distorted.

The first striking modifications which such an approach will introduce into the "traditional" picture will be the need for the conceptualization of four different kinds of vectors, only the first of which can be fruitfully represented by an arrow. Although this may seem at first like an unnecessary complication, we shall find that, as a result, the degree to which our geometrical intuition can be harnessed for an understanding of the subject will be enormously enhanced. We return to this point in Sec. 1.7.

And so we begin at the beginning.

1.3 What Vectors Are Not

A definition of "vector" which is often encountered in elementary physics is that it is *something possessing both magnitude and direction.* Taken literally, this would include an automobile − which does, after all, possess both magnitude and direction; and even though no thinking person would actually make such a mistake, it does point out that, in fact, the definition is empty. The problem, in other words, is not that someone might confuse an automobile with a vector, but that the basic properties of a vector have in no way been clarified.

A rather more treacherous definition − in the sense that it is not merely a joke, but contains a serious half-truth − is that *a vector is a set of three numbers, called components;* or somewhat more formally, *a vector is a set of numbers carrying a single index that runs from 1 to the dimensionality of the space.* We say that this is "a half-truth" because there do exist applications, particularly those involving matrix algebra, in which such a nomenclature is conveniently and validly used. Nonetheless, the concept defined in that way lacks some fundamental properties that we ordinarily like to take for granted.

Consider, for example, a "vector" drawn to characterize a person, with the three components representing that person's age, height, and weight; Fig. 1.1a shows such a construction as it would pertain to the author of this book (at the time of writing). What we see here are three appropriately labeled axes, and an arrow that signifies the complete "vector." But if, as in Fig. 1.1b, we erase the three axes leaving only the arrow, this arrow becomes totally meaningless. Nothing in particular happens in the direction that the arrow points; nor is anything signified by its size.

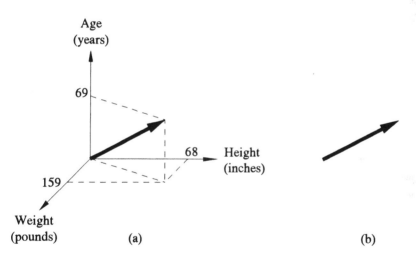

(a) (b)

Fig. 1.1: A "vector" which depends on components

For a true vector, as we shall soon see, the logical connection between the "arrow" and its components is reversed, in that the arrow itself carries the fundamental meaning. True, one can define a "coordinate system," or "reference frame," which then implies numerical values for the vector's components, but the same vector will have different components in different systems. This last statement emphasizes — by using the phrase *the same vector* — that the identity of the vector does not depend on any particular system of axes.

1.4 What Traditional Vectors Are

From the discussion of the previous section, we are led to the following definition: *A vector is a quantity which is usefully represented by an arrow.* A familiar example is the velocity of a moving particle: the direction of the arrow then corresponds to the direction in which the particle is moving, and this correspondence has nothing to do with any particular coordinate system. In the same way, the length of the arrow gives the distance covered by the particle in unit time. Notice that this, too, does not depend on the coordinate system, even when the scale of the axes is modified: if, for example, we change from feet to meters, the length of a given arrow will be numerically smaller, but so also will be the numerical distance which the particle covers in unit time. (Note that the unit of time *does* make a difference, but time is not part of the geometry of our three-dimensional space.)

   To state the above properties somewhat differently, the correspondence between the physical quantity (velocity) and the mathematical object (arrow) is preserved (a) if we arbitrarily rotate our coordinate axes, (b) if we arbitrarily change the numerical scale of our coordinate axes. Traditionally, however, only the first of these is required to define a vector.

   Consider, for example, the physical situation of a parallel-plate capacitor whose plates are very large and 1 cm apart, with a potential difference of 30 V maintained between them. Let the vector that goes from the negative to the positive plate, and is perpendicular to them, be denoted by d; let the electric field vector between the plates (of magnitude 30 V/cm) be E (Fig. 1.2). Now if the coordinate axes are rotated, the correspondence between those quantities and the arrows that represent them will be preserved; but if we alter the numerical scale, this may no longer be so. To illustrate, if we change from centimeters to meters, the numerical value of d will go from 1 to 0.01; but the numerical value of E will go from 30 (V/cm) to 3000 (V/m). Thus the relation of arrow to quantity will be maintained for d but not for E. (The unit of potential is, like the unit of time, not part of our geometry and ought consequently not to change; but see Problem 1.1.)

In the traditional approach we ignore this difficulty, making peace with the fact that the equivalence between vectors and their geometrical representations (that is, arrows) is maintained for rotations of coordinates but not for other transformations. In the present treatment, by contrast, we wish to be more ambitious,

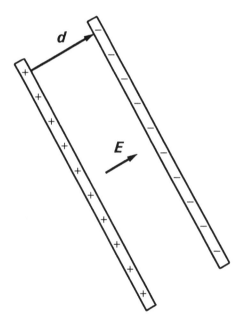

Fig. 1.2: Two vectors which behave differently

and look for a formulation that does not have such a shortcoming. What the preceding example has made clear is that, in order to be able to deal with more general coordinate transformations (of which even the above "change of scale" is still only a very special example), *arrows alone will not suffice.*

1.5 Do Vectors Have a Location?

In constructing Fig. 1.2, we defined **d** as going from one capacitor plate to the other; yet, somewhat paradoxically, it can be drawn equally well someplace else, as in Fig. 1.3. In this case, the same defining statement simply means that *if* **d** is slid over so that its tail is on the negative plate, *then* its head will fall on the positive plate. This illustrates how our vector concept

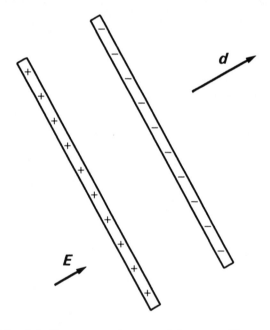

Fig. 1.3: Vectors may be moved about at will

allows an arrow to be slid around as much as we please without changing its value. When this rule of movability is (as in the same figure) applied to the electric field, it looks even more confusing, since the **E**-arrow now finds itself in a place where the actual electric field vanishes. No matter; we must simply remember that *E* represents *the field in the center of the capacitor*, regardless of where it is drawn.

There are, of course, situations in which one is interested, not merely in the electric field at a single point, but in the way that it varies from point to point. In that case, one needs to define *E* as a function *E(r)*, where *r* is a *radius vector* that tells us the point's location. This does not mean, however, that the *E(r)* for each *r* must be drawn *at* the corresponding *r*; in fact, that would be impossible since *r* denotes a geometrical point whereas the arrow *E(r)* necessarily has a finite size.

In general, regions of space at each point of which a vector quantity is defined are called *vector fields,* and we shall take up their detailed study in Chapter 5.

1.6 Coordinate Transformations vs *Distortions of System*

The capacitor example of the last section illustrates how a physical quantity is determined by the physical experiment that engendered it; so, for instance, the electric field *E* was produced by a certain configuration of two conducting plates connected to a certain battery. In discussing "coordinate transformations," such as rotation or change of scale, we assumed that the experimental setup remains untouched, only the reference frame (which is, after all, a figment of our imagination) undergoing any modification.

There is, however, another, closely related way of discussing the subject, and that is by imagining that it is not the reference frame but the physical setup which is changed. For example, the correspondence under rotation between the arrows *d* and *E* in Fig. 1.2 and the respective physical quantities is investigated by imagining that it is the plates themselves that are rotated, rotating the diagram, including the arrows that it contains, with them. It is clear that if this is done, the *E*-arrow will still correspond correctly to the (new) electric field, as will *d*, of course.

For the second type of transformation − change of scale −

we proceed in a corresponding way: we keep the reference scale the same (that is, we continue to use centimeters), but "squeeze" the apparatus down to a smaller size. (Naturally, no demand is placed on any physical elasticity of the capacitor plates: "squeezing" the apparatus simply means rebuilding it on a smaller scale.) By definition, such a compression of the system must be accompanied by a compression of the diagram and of the arrows that appear in it. In this way we reach the same conclusion as before: the compressed arrow d continues to correspond to the spacing between the plates, but the compressed arrow E has lost its reference: it, too, has become smaller, whereas the actual electric field produced by the same potential difference acting across a smaller gap has actually grown bigger! Thus our conclusion that, for transformations more general than mere rotations, arrows alone will not be enough, holds equally well whether we think of those transformations as acting on the coordinates or as distorting the physical system.

1.7 Why Is Topological Invariance Important?

In Sec. 1.2, we mentioned a desire to formulate all concepts and relations in ways which do not require the numerical measurement of distances or angles; which are, in other words, *topologically invariant.* The ability to make such formulations is important to us on two levels, the intuitive and the practical.

Intuitive: Much of our perception of three-dimensional space takes place through images cast on a two-dimensional retina, so that (over and above permanent distortions of the eye to which the brain presumably accommodates) the pattern we see is subject to ever-shifting distortions as one's point of view changes; the visual size of an object, for example, seldom remains constant for long. Yet vision (or, for that matter, other senses) would not be of much use to us unless our minds had long ago learned to abstract from all such images and "see" the underlying reality, by developing an ability to interpret what we see independently of distortions of both distances and angles that occur in the process of perception. To formulate physical reality in topologically invariant terms is, thus, to begin with an intuitive capacity that we already possess and to develop it further.

Practical: Ultimately, what we now describe as distortion of space will turn into transformation of coordinates; if one changes, for example, from Cartesian to spherical coordinates, the structure of a physical system relative to the coordinate surfaces becomes radically different. However, any operation that we have succeeded in defining topologically will, automatically, maintain its algebraic form independently of the coordinate system in which we choose to work, obviously an enormous convenience.

Of course, not all laws of physics are purely topological in nature, as we shall have occasion to emphasize again in Chapter 8; indeed, in a certain sense it may be said that the non-topological laws are where the "real" physics resides. One way or the other, however, the ability to distinguish one type from the other will be of great help in clarifying both their meaning and their rôle in overall physical theory.

PROBLEMS

1.1   In Sec. 1.4, we discussed the behavior of E under a change of scale by assuming that the potential difference between the plates remains constant, as though they were connected to a battery. Suppose we assume, instead, that the plates are "floating," so that it is their *electric charge* that remains the same. Will the behavior of E now be different? Will it be more like an arrow?

1.2   Suppose that the spacing between plates is kept constant but they are laterally compressed by a certain factor. Will the arrow d change its value? Does the correspondence between arrow and physical value persist?

1.3   Repeat the previous problem considering E, for the two assumptions of (a) constant potential difference and (b) constant charge. In each case, will E

change? Will it behave like an arrow?

1.4   How many physical quantities can you think of that can, like d, be usefully represented by an arrow?

2

VECTOR TYPES AND
VECTOR OPERATIONS

2.1 What Is the Problem?

The explorations of the previous chapter confirmed that traditional vectors are "like" arrows in that they do, indeed, behave like arrows under rigid rotations. But it turned out also that in the case of some, *but not all,* traditional vectors the correspondence goes considerably further. In particular, there appear to exist some vectors which behave like arrows, not just under rigid rotations, but under many more − perhaps all? − types of transformations.

The prototype of this latter type of vector is a *displacement,* that is, an operation that takes a certain point from "here" to "there," since it can be put into a natural correspondence with an arrow which begins "here" and ends "there" (Fig. 2.1). Such an identification makes it clear that if space is, for example, compressed in some direction, the arrow gets shorter in exactly the same way in which the point "there" becomes closer to the point "here," so that the correspondence remains valid. We express this fact by saying that displacement is not merely a *vector* but an *arrow vector,* or an *arrow* for short.

Since, in our intuitive three-dimensional world, time is not related to the dimensions of the space and hence is best thought of as remaining the same when space undergoes arbitrary

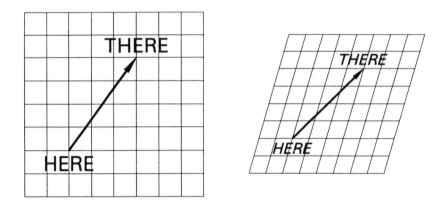

Fig. 2.1: A displacement compresses like an arrow

transformations, it follows that any quantity related to displacement by a factor of time will also be validly represented by an arrow. To this category belongs the *velocity* of a particle, which we already mentioned in Sec. 1.4; it is defined as the displacement that the particle undergoes in a fixed (usually infinitesimal) time Δt, divided by Δt. By observing what the particle does in such an interval, we can again identify a "here" and a "there" between which it travels, and so construct a velocity arrow. Needless to say, *acceleration* acts similarly.

On the other hand, the explorations of the previous chapter taught us also that not all traditional vectors behave in this way; for example, the electric field *E* of Sec. 1.4 does not. We are thus led to ask whether there may exist other possible geometrical representations of vectors, which behave under rigid rotations just as arrows do, but may correspond better to some physical quantities when the discussion is widened to include more general transformations. This is the question that will occupy us in the present chapter.

2.2 Stacks

As our first example of a geometrical object which is a vector and yet not an arrow, we introduce the concept of a "stack." As long as we limit ourselves to rigid rotations, stacks and arrows have the same properties; in fact, it is possible (as we shall see) to define a one-to-one correspondence between the two. But as soon as we go on to more general transformations, the two no longer behave in the same way.

First, the definition: A stack consists of a number of parallel sheets, plus a loose arrowhead to indicate the "sense" (Fig. 2.2). The direction of the stack is defined by the *orientation* of the

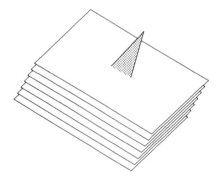

Fig. 2.2: A stack

sheets, whereas its numerical magnitude is given by their *density*. (Recall that, for an arrow, direction is defined by the orientation of its shaft, and magnitude by its *length*.)

It is immaterial how many sheets of the stack are shown, or how large they are drawn, just as it is immaterial how thick the line of an arrow is made. In principle, two sheets (plus the arrowhead) are sufficient, though it is often less confusing to sketch in a few more.

The one-to-one correspondence between stacks and arrows, which can only be invoked if one limits oneself to rigid rotations, is that to each stack there corresponds an arrow whose direction is perpendicular to the sheets of the stack, and whose length is equal to the density of those sheets, that is, to the reciprocal of the spacing between them. But as soon as we use the word "perpendicular" we know that we are restricted to rigid rotations, because more general distortions of space will not preserve the condition of perpendicularity, and the supposed one-to-one correspondence breaks down; furthermore, equating a length to a density requires a unit of length, which will also not remain the same. For such general transformations, arrows and stacks are *really different* from each other.

A particularly simple contrast becomes apparent if we construct the arrow which supposedly "corresponds" to a given stack, and then compress space along the direction of that arrow. Clearly, this will cause the arrow to become smaller (that is, its length is diminished), at the same time causing the stack to get larger (that is, its density is increased), so that it is impossible for the correspondence to be maintained. We recognize immediately that the behavior of the electric field E, as we discussed it in Sec. 1.4, identifies it as not an arrow vector but a stack vector.

2.3 The Size of Mathematical Objects

We have talked about how arrows and stacks are modified when the space in which they exist is distorted by "arbitrary" transformations, but have until now ignored the possibility that this distortion makes the arrow into something that is no longer an arrow, for example by curving it; or perhaps makes the spacing of the stack sheets non-uniform or bends them into surfaces that are no longer plane, so that the stack is no longer a stack. As long, however, as the transformation is continuous and smooth, so that our space is neither "torn" nor "creased" by it, this type of problem can be made negligible to any desired degree simply by drawing our stacks and arrows *small.*

There is no real loss of generality in such a requirement, since the constant of proportionality that relates the geometrical length of our arrow, or the density of our stack, to its physical counterpart is, after all, arbitrary. So, for example, there is nothing to prevent us from representing the velocity of a particle moving with the speed of light by an arrow which is one micron long; or some very weak electric field by a stack whose numerical density is very high, so that the two sheets that constitute its minimum representation can be squeezed into a tiny space. (An exception is the displacement vector itself, whose size is locked to the scale of the coordinate system: it would make little sense, for example, to denote the displacement from Ann Arbor to Chicago by an arrow which is "small." For this reason, only *infinitesimal* displacements — like the ones involved in the definition of velocity — can, strictly speaking, be represented by arrows under arbitrary, and not just linear, space trans-formations.)

The ability to draw symbols on a scale that is small is what allows us to use the word "arbitrary" in referring to the transformations under which the behavior of our vector operations remains invariant, and to describe all resulting relationships as "topological." The only restriction that does have to be respected is that the transformations be *continuous and differentiable;* and we can fully expect that at singular locations (for example, the polar axis in the case of a transformation from Cartesian to cylindrical coordinates) some of the quantities in which we are interested will exhibit anomalous behavior. However, although phenomena arising from singularities in a coordinate system can under some circumstances be quite important, they are always (in physical applications) limited to special points or lines. Thus it remains appropriate to assume that at general points of our space they do not occur, and to postpone their consideration to situations where both their origin and their consequences can be examined in a more specific manner.

2.4 Orientation of Lines and of Planes

Although the directions of arrows and stacks are both determined by their respective orientations, the meaning of "orientation" is rather different in the two cases, since in the first we are referring to the orientation of a line, and in the second to the orientation of a plane. So, for example, we can say that two arrows have "the same" direction, meaning that their shafts are parallel; but there can be no corresponding meaning to the claim that a certain arrow and a certain stack have directions which are "the same," since the two types of orientation refer to different qualities. It is true, of course, that classical solid geometry allows one to apply the term "parallel" in all three cases: a line can be parallel to another line, a plane can be parallel to another plane, and a line can be "parallel" to a plane. But such usage is in reality no more than a confusing pun. In the first two cases the property of being parallel determines a unique orientation: two lines, each parallel to a given line, must be parallel to each other, and similarly for planes. But two lines, each "parallel" to a given plane, can have quite different orientations; and the same is true for planes which are "parallel to a given line."

At the same time, we note that the relation between a line and a plane which classical geometry describes as being "parallel" does have a particular importance to us since it is, as a matter of fact, topological; if, for example, the shaft of an arrow vector is contained in one of the sheets of a stack vector, that fact will not change if space is transformed in an arbitrary manner. This situation faces us with a terminological problem. In the traditional setting, in which the stack is "really" an arrow directed perpendicularly to the stack, we would describe the relation between an arrow and a stack in one of whose sheets it is contained as being "perpendicular" to each other; but to us such a name is unattractive, both because it "smells" of being metrical, and because describing a line which is contained in a sheet as perpendicular to that sheet obviously invites confusion. On the other hand, we do not wish to use "parallel" for this relation, because it not only lacks the uniqueness property of parallelism but radically contradicts the traditional ideas of vectors.

Accordingly, in this book we shall simply say, for the case in which an arrow vector is parallel to the sheets of a stack vector, that the arrow is *contained in* the stack; or equivalently, although less familiarly, that the stack is *contained in* the arrow. Our hope is that the cost of using a somewhat unusual term will be compensated by the confusion which such usage avoids.

To recapitulate: in the language that we adopt, a pair of arrows can be parallel to each other (but not perpendicular), and the same is true of a pair of stacks; but a stack and an arrow *cannot* be parallel *or* perpendicular. An arrow and a stack can be *contained* one in the other, but not two arrows or two stacks. The reader will, of course, understand that when we use the expressions "can be" and "cannot be," we are simply stating what restrictions exist if we insist on purely topological relations – which, however, we do, for the present at least.

2.5 Algebra of Arrows and of Stacks

The simplest algebraic operation that can be performed on a vector is multiplication by a scalar, since it follows directly from the definition of "magnitude." Thus, to multiply by a scalar c, we multiply the length of the arrow, or the density of the stack, by the magnitude of c; in the latter case, it is the same as *dividing* the spacing between the sheets by the same number. If c is negative, we reverse the sense of the vector as well. It is worth remarking that whereas one vector can perfectly well be the negative of another, meaning that one is obtained from the other by multiplication by -1, it does not follow (as it would for scalars) that one of them is positive and the other negative; in fact, there is no such thing as a "positive" or "negative" vector. The relationship of being another vector's negative is, in other words, unlike the case of scalars, completely reciprocal; it does not put either vector into any particular class. We shall have occasion later to return to this point.

The next most fundamental operation is, of course, addition. For the case of arrows, its definition is embodied in the familiar "parallelogram rule": one draws the two vectors with their tails together, completes the parallelogram, and draws a third vector from the common tail of the original two vectors to the point at which the parallelogram was completed. This definition is so familiar that most people have never asked themselves why it makes sense to call it "addition," a word which is, after all, already defined for scalars. The reason is that, as one can easily verify, this definition satisfies the three fundamental properties of addition, namely the commutative law

$$A+B=B+A \; ; \tag{2.5.1}$$

the associative law

$$A+(B+C)=(A+B)+C \; ; \tag{2.5.2}$$

and the distributive law with scalar multiplication

$$a(A+B)=aA+aB. \tag{2.5.3}$$

For stacks, addition is defined by a somewhat similar parallelogram rule, as follows. If we superimpose the two stacks on top of each other, they will form a kind of "honeycomb" the cross sections of whose "cells" are parallelograms; we then construct a new stack whose sheets are the diagonal planes of these (Fig. 2.3). There are, of course, two such sets of planes, corresponding to the two diagonals which a parallelogram has, so an additional specification is required. It consists of associating each side of the parallelogram with the arrowhead corresponding to the stack of which it is a part, and then drawing the diagonal so that one of the resulting triangles has both original arrowheads pointing *into* the triangle.

Having thus identified the correct set of diagonal planes, we attach an arrowhead to them such that it points *out of* the triangle for which the two original arrowheads point *inward;* the new stack is, by definition, the *sum* of the two original ones. Again, it is straightforward to verify that the three fundamental laws of addition are satisfied by this definition.

The construction of Fig. 2.3 is, of course, impossible if the two stacks are parallel to each other, but in that case we simply

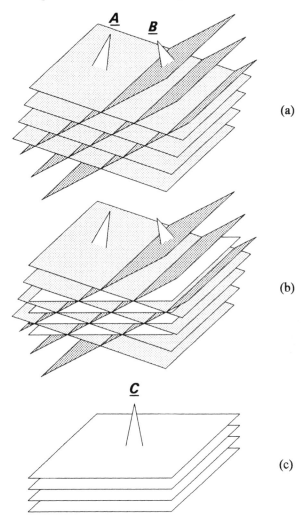

(a)

(b)

(c)

Fig. 2.3: Addition of two stacks

construct the sum stack as also having the same direction, and having a magnitude which is the sum of the original ones (or, if their senses are opposite, the difference). It is satisfying to know that the same result is obtained if we consider two stacks whose directions are *almost* the same, and examine the limit which is approached as those directions approach each other (Problem 2.6).

We emphasize once more that the rules for multiplying an arrow or a stack by a scalar, as well as those for adding two arrows or two stacks, are entirely topological; that is, they do not depend on using a particular length scale, or on measuring any angles. On the other hand, the addition of a stack and an arrow remains undefined. That is not to say, however, that other operations that combine those two types of vectors don't exist; but before going on to investigate them, we pause to define a bit of notation, as follows. When assigning a symbol to an arrow or to a stack, we shall place an arrow over a letter to indicate an arrow vector, and underline it to indicate a stack vector, as follows:

$$\vec{R} \text{ is an arrow; } \underline{K} \text{ is a stack.} \qquad (2.5.4)$$

As with other notational conventions, it is not necessary to consider this a commitment for life; it will, however, help us to clarify our discussions as long as the underlying concepts are not yet completely familiar.

2.6 The Dot Product

The traditional definition of the dot product of two vectors is well known (Sec. 1.2): it is a scalar, equal to the length of one of the vectors multiplied by the projection upon it of the other vector; or, somewhat differently, it is the product of the magnitudes of the two vectors multiplied by the cosine of the angle between them. Quite obviously, such a definition (stated here in terms of arrows, since arrows are the traditional way in which vectors are represented) cannot possibly be invariant to a general coordinate transformation, since it involves measurement both of angles and of numerical lengths. Yet a different definition, equivalent to the traditional one if one restricts oneself to rigid rotations, can be constructed, provided that *a dot product is always taken of one arrow and one stack*. The definition is the following: We move the arrow onto the stack and count the number of stack sheets which the arrow spans. The dot product is positive if the arrow spans the sheets in the order of the arrowhead attached to the stack, and negative otherwise. An example of this construction, in which $\underline{K} \cdot \vec{R}$ is equal to $+7$, is shown in Fig. 2.4.

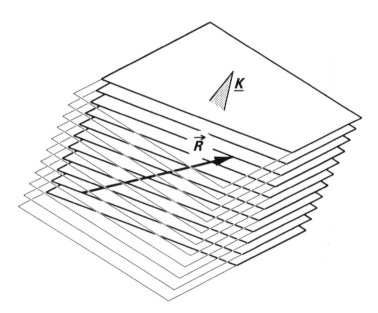

Fig. 2.4: Dot product of arrow and stack

We write the dot product of $\vec{R}$ and $\underline{K}$ as $\underline{K} \cdot \vec{R}$ or, equivalently, as $\vec{R} \cdot \underline{K}$; in other words, it is the same whether the arrow or the stack appears first in the product. Thus the dot product is commutative by definition, which need not be true of other types of products; in fact, the cross product, which will be introduced in the next chapter, does not have this property.

The striking characteristic of our new definition is, of course, its invariance under coordinate transformations or, what here amounts to the same thing, distortions of space: if, for example, the tail of our arrow is anchored in a certain sheet of the stack, and its head in another sheet, this will remain true even after an arbitrary deformation has occurred. Thus the numerical value of a dot product, having been defined in a purely topological way, becomes invariant to the most general coordinate transformations.

Incidentally, one can also verify that the distributive law of multiplication holds for the dot product as here defined; that is,

$$(a\vec{R}+b\vec{S})\cdot(c\underline{K}+d\underline{L}) =$$

$$(ac)(\vec{R}\cdot\underline{K})+(ad)(\vec{R}\cdot\underline{L})+(bc)(\vec{S}\cdot\underline{K})+(bd)(\vec{S}\cdot\underline{L}) . \qquad (2.6.1)$$

PROBLEMS

2.1   In Sec. 2.2, we concluded that the field E of Fig. 1.2 is a stack, but we did this on the assumption that, as space is compressed, the potential difference between the plates remains constant. Suppose instead that it is the charge on the plates that remains constant; is a stack still a good representation of E? Would an arrow be better?

2.2   Referring again to Fig. 1.2, we said that the magnitude of d is 1 (cm), and the magnitude of E is 30 (V/cm). By what factor does space need to be expanded for the two to become numerically equal?

2.3   Find the dot product $\vec{d}\cdot\underline{E}$ both before and after the expansion of the previous problem.

2.4   In a traveling plane wave, we speak of a "propagation vector" k such that $k\cdot\vec{\Delta r}$ is the phase difference between two points separated by a vector $\vec{\Delta r}$. Do you think this phase difference ought to be the same when the coordinate system is compressed? Why? What flavor of vector would you use for k?

2.5   Referring to Fig. 2.3, imagine the operation of addition in a space in which we are allowed to use rulers and protractors; and that in this space the stacks $\underline{A}$ and $\underline{B}$ have magnitudes A and B, with an angle θ between them. Prove that the square of the magnitude of their sum is

$$A^2 + B^2 + 2AB \cos\theta \,,$$

just as would be true for arrows.

2.6   Prove that if the construction of Fig. 2.3 is applied to two stacks whose directions approach each other, the limit of the sum is a stack of that same direction, with a magnitude equal to the sum of the two original magnitudes.

2.7   Again imagine, as in Problem 2.5, a space in which we are allowed to use rulers and protractors. Show that the dot product $\underline{K} \cdot \vec{R}$ of Fig. 2.4 is equal to the magnitude of $\underline{K}$ times the projection of $\vec{R}$ onto a line perpendicular to the sheets of $\underline{K}$.

3

MORE OPERATIONS, MORE VECTORS

3.1 A New Kind of Product

In the previous chapter we discussed two types of vectors: arrows and stacks. For each one, we defined two operations − multiplication by a scalar and addition − using definitions which are entirely topological, that is, do not depend on numerical measurement of lengths or angles. We also defined one operation, the dot product, which necessarily involves one arrow and one stack; again, the definition is topological, so that the numerical value of a dot product is unchanged when space is transformed in an arbitrary way.

In this chapter we shall define a new operation, called the *cross product*, whose nature is in a way the opposite of a dot product: a cross product can be taken between two stacks or between two arrows but *not* between a stack and an arrow. This operation does, however, cause some new complications, beginning with the fact that the cross product of two arrows, although itself a vector, is neither an arrow nor a stack; nor is the cross product of two stacks. In other words, the introduction of this new kind of product also forces us to introduce some new kinds of vectors. A second complication, to be addressed later in the chapter, is that the concept of the cross product forces us to reexamine what exactly we mean by the "sense" of a vector.

3.2 The Cross Product of Arrows

In the traditional approach (Sec. 1.2), the cross product of two vectors — that is, arrows — is defined as a new arrow whose direction is perpendicular to both of the original arrows, and whose magnitude is equal to the area of the parallelogram subtended by the two of them. It is immediately clear, however, that an object defined in this way is not invariant to general transformations: even in a simple change of scale, for example, the new quantity will vary with the *square* of the factor by which distances are compressed. And, of course, we know that the property of perpendicularity is not topologically invariant either.

What we do instead is to take the bull by the horns and define a third type of vector, which we call a *thumbtack*. A thumbtack, whose algebraic symbol we will take to be a boldface letter with

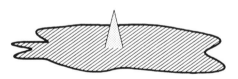

Fig. 3.1: A thumbtack

a small circle under it like this: $T_\circ$, is a single finite piece of plane, with a sense indicated by a loose arrowhead (Fig. 3.1). The direction *type* of a thumbtack is obviously that of a stack, and not that of an arrow. In other words, it is meaningful to say that a thumbtack is parallel to a stack, but not to an arrow; it can, however, be *contained in* an arrow, but not in a stack (Sec. 2.4).

We define the magnitude of a thumbtack to be its area, the shape being immaterial. Two thumbtacks are, in other words,

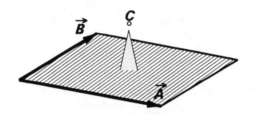

Fig. 3.2: The cross product of two arrows is a thumbtack

considered identical if they are parallel to each other and their areas and senses are the same, even though one may be, say, circular and the other square. In addition, the thumbtack has its own algebra: it can be multiplied by a scalar, and any two of them can be added, according to definitions which are entirely topological, and which follow the commutative, associative, and distributive laws of algebra. We will discuss these in detail in Sec. 3.7, after the fourth (and final) type of vector has been introduced.

It now becomes simple to define the cross product of two arrows as *a thumbtack determined by the parallelogram which those two arrows subtend* (Fig. 3.2). It is, in other words, *the thumbtack which is contained in each of those arrows, with a magnitude equal to the area of the parallelogram subtended by them.* It follows from this definition that the cross product of two parallel arrows is zero.

3.3 Polar and Axial Sense Genders

Traditionally, the sense of a cross product is specified by a right-hand rule; and we can, if we wish, accept the same idea for defining the sense of our new thumbtack. In particular, to obtain the sense of the cross product $\underset{\circ}{C}=\vec{A}\times\vec{B}$, one first moves $\vec{A}$ and $\vec{B}$ with their tails together, then holds one's right hand in a very loose fist so that the fingers are curled *from* $\vec{A}$ *to* $\vec{B}$; the direction of the thumb is then the direction to be assigned to $\underset{\circ}{C}$ (that, in fact, is how Fig. 3.2 was constructed). A definition based on this rule remains invariant for any transformation which does not change a right hand into a left hand; or equivalently, for any distortion of space which can be produced continuously from its original condition.

There do, however, exist space transformations which can *not* be generated by such a continuous process. This is illustrated in Fig. 3.3, which reproduces Fig. 3.2 (on the right) while also depicting (on the left) its reflection in a vertical mirror (indicated by the double line). The reader is invited to verify that the

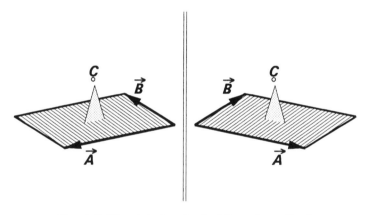

Fig. 3.3: The right-hand rule fails under reflection

"sense" of $\underset{\circ}{C}$ on the left side of the mirror, although a faithful reflection of the one on the right, follows the exact opposite of the right-hand rule we just formulated.

We thus see that insisting on the use of a right-hand rule raises a serious problem. Of course, one way of dealing with it is simply to make peace with the fact that cross products are not invariant under spatial reflections; that still leaves us a wide range of invariance, sufficient for a great many purposes.

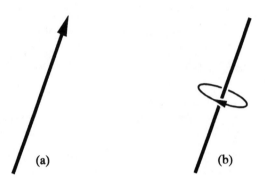

Fig. 3.4: Polar and axial senses of an arrow

Yet there are important physical problems where it is useful not to be limited in that way. We then have available a more ambitious option, and that is to note that the concept which we call "the sense" itself comes in two "genders." For an arrow, for example, we can put an arrowhead at one end (Fig. 3.4a); or else

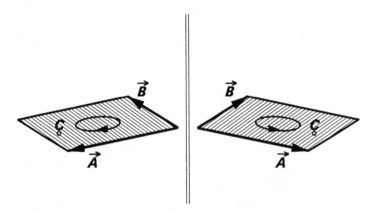

Fig. 3.5: The axial sense works under reflection

a "curlicue" around the segment, as though it were rotating about itself (Fig. 3.4b). We refer to the first as a *polar sense,* and to the second as an *axial sense.* The penalty paid for taking this option is, of course, that the number of vector species is doubled; it should be understood, however, that the doubling is not a quirk of our conventions but something built into the nature of physical interactions.

The problem of defining the sense of a cross product so as to be invariant under reflections is solved if we are willing to say that $\vec{A} \times \vec{B}$ is not a polar but an axial thumbtack; that is, its sense is not determined by an arrowhead but by a curlicue drawn within it. We then specify the particular direction of that curlicue by requiring that it correspond to the direction in which $\vec{A}$ needs to be rotated so as to align it with $\vec{B}$. Such a construction is shown in Fig. 3.5, from which we see also that it maintains its validity even when reflected in a mirror.

We note that, whether one chooses axial senses or right-hand rules, the sense of a cross product is always reversed when the order of its factors is reversed.

3.4 Algebra of Polar and Axial Senses

The extension to this situation of the algebraic laws we have previously introduced is straightforward; thus, the sum of two polar arrows is a polar arrow (as before), and the sum of two axial arrows is an axial arrow (according to conventions which the reader will have no difficulty developing). By contrast, the sum of one axial and one polar arrow must remain undefined.

The cross product of two polar arrows is, as we saw, an axial thumbtack. The cross product of two axial arrows is also an axial thumbtack, but its construction is a bit different. First, we bring the two arrow shafts together so that, viewed from the common point, their curlicues are in the same direction (either clockwise or counterclockwise, it makes no difference which). The axial direction of the cross product is then again the direction that takes the first factor into the second.

Finally, the cross product of one polar and one axial arrow vector is a polar thumbtack, which is constructed as follows. First, we may assume that the first factor is the axial one, since if it is the other way around we need only change the order of the factors, knowing that in so doing we reverse the sense of the product. We then form the parallelogram of the two, attaching *either* end of the axial vector to the *tail* of the polar vector. Having done this, we find that the curlicue that defines the sense of the first factor *pierces the parallelogram* in a certain direction; that direction then defines the sense of the cross product.

Although it would obviously be possible to elaborate our notation further so as to be able to indicate the polar/axial gender of vectors and scalars directly, we shall not do so in this book, feeling that it would complicate the symbol set more than is appropriate. Accordingly, the reader must keep in mind that symbols such as $\vec{A}$ and $\underset{\circ}{B}$ can represent either polar or axial quantities. If, on the other hand, we are dealing with a problem in which reflection invariance is not important, we may consider

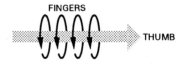

Fig. 3.6: Universal right-hand rule

the two sense genders to be equivalent to each other if they are related by the *universal right-hand rule* which is illustrated in Fig. 3.6. We obtain this figure by holding up our right hand, with the thumb pointing to the right, and allowing the four fingers to curl loosely. The axial sense indicated by those fingers is then equated to the polar sense of the pointing thumb.

3.5 Polar and Axial Scalars

At this point we need to go back and reexamine the dot product, which was defined in Sec. 2.6 as involving one stack and one arrow. The result, we said there, is a scalar, whose sense (that is, sign; for a scalar, the two are synonymous) is positive if the arrow spans the sheets of the stack in the direction of the stack, and negative if those two are opposite. Such a definition is valid if the stack and the arrow are both polar or both axial; but if one of them is axial and the other polar, their dot product is a new entity called an *axial scalar* (more commonly called a *pseudoscalar*). The axial scalar is rather confusing to define, since the "sense" of an ordinary (that is, polar) scalar is simply + or − (unless, of course, it is zero); in other words, a polar scalar is either positive or negative. By contrast, an axial scalar also has one of two senses, *but they are neither + nor −*.

It is useful to invent two new symbols, $\circlearrowright$ and $\circlearrowleft$, to denote those two senses, and to use the terms *right-handed* and *left-handed* to describe them. The (axial) sense of the dot product of an arrow and a stack, one of which is polar and the other axial, is then specified by the appearance of the curlicue of the axial one when viewed along the direction of the polar one. Although these two senses are opposites of each other, so that we can say that $-(\circlearrowright 5) = \circlearrowleft 5$ or $\circlearrowright 5 = -(\circlearrowleft 5)$, the new symbols individually have no one-to-one correspondence with + and −; so, for example, it is impossible to say whether $\circlearrowright 5$ is *greater than* $\circlearrowleft 7$ or vice versa. But confusing as this may seem, it should also be remembered that in this respect it is ordinary (polar) scalars, and not axial ones, that form the exception among types of physical quantities. For vectors too, after all, a statement like $-\vec{A} = \vec{B}$ is perfectly meaningful even though neither $\vec{A}$ nor $\vec{B}$ can be said individually to be either "positive" or "negative."

The reader may wonder why one cannot compare two axial scalars like $\circlearrowright 5$ and $\circlearrowleft 7$ by comparing their *magnitudes*. The answer is that indeed one can, but that the meanings of the two types of comparison are quite different − as is also true for ordinary scalars. Suppose, for example, that we number the floors of a building ..., $-3, -2, -1, 0, 1, 2, 3, \ldots$ using 0 for street level and negative numbers for basements (as is often done in Europe). The statement $2 > -3$ then signifies that the second floor is higher than the third basement; at the same time, the magnitude of -3 is greater than the magnitude of 2, signifying that the third basement is further from street level than the second floor. Clearly, the two give answers to two different questions.

Of course in the case of a pair of vectors even a comparison of magnitudes becomes questionable because, unless they are parallel, it is possible for space to be compressed in such a way that the vector that was previously larger now becomes smaller.

In physical applications, there are certain scalar quantities, such as energy or entropy, for which the relations *greater than* and *less than* possess an absolute meaning: so, for example, a radiating system will always go toward *lower* energy, and a closed thermodynamic system toward *higher* entropy. This tells us that any mathematical formula that purports to compute an energy or an entropy must either produce a result which is a polar (and not an axial) scalar, or else it describes a process (such as beta decay) which is *physically* not invariant under mirror reflection.

3.6 The Fourth and Last Vector: The Sheaf

We said earlier that unlike the dot product, which exists only between vectors of opposite "flavor," the cross product requires the two to be the same, implying that not only two arrows, but two stacks may also have a cross product. That is, indeed, the case; and to understand it will require us to define a fourth (and last) type of vector, which we call a "sheaf."

Consider two stacks superimposed on each other (Fig. 3.7). The intersections of the sheets belonging to one with the sheets belonging to the other form a family of lines, whose *density* is proportional to the magnitude (that is, sheet density) of each of the stacks, when their directions are kept constant. If, on the other hand, their directions are varied while their magnitudes remain fixed, the density of intersection lines goes to zero if the two stacks are made parallel to each other, and reaches a maximum when their directions are perpendicular. Both of these properties suggest that we are here, in fact, dealing with *the cross product of the two stacks*.

To specify it precisely, we define a new type of vector, called a "sheaf," as a bundle of lines whose *density* indicates the vector's magnitude, and whose direction is the direction of those lines. Just as is the case with the sheets that make up a stack, the exact pattern in which we distribute the lines of a sheaf is unimportant; only their density matters. So, for example, we can

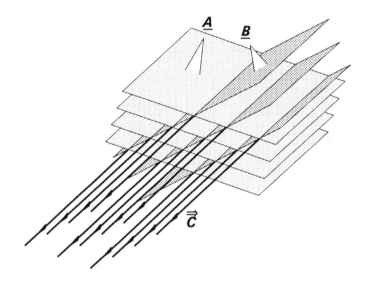

Fig. 3.7: Cross product of two stacks

halve the spacing between them in one direction while doubling it in the other, and end up with a sheaf that is by definition identical.

As for the sense, we have, as usual, two choices. Either we indicate it by arrowheads on the lines, as in Fig. 3.7 ("polar sheaf"), and define the sense of a cross product of two polar stacks by a right-hand rule that relates it to the senses of the two factor stacks; or we admit the two genders of sense — polar and axial — that we have already discussed, and make the cross product of two polar stacks into an axial sheaf whose sense requires no right-hand rule for its definition. In this latter case, the cross product of two axial stacks would also be an axial sheaf, whereas the cross product of one axial and one polar stack would need to be defined as a *polar* sheaf.

We denote a sheaf vector by a boldface letter with a double arrow over it, like this: $\overset{\Rightarrow}{\boldsymbol{J}}$.

3.7 Algebra of Thumbtacks and Sheaves

Thumbtacks and sheaves have rules of algebra entirely analogous to those of arrows and stacks. The simplest of these is, as always, multiplication by a scalar, and it is (as one might imagine) performed by identifying the quantity which signifies the *magnitude* of the vector and multiplying it by the given scalar, leaving the direction of the vector the same, and reversing the sense if the multiplier is negative.

For the thumbtack, the magnitude is defined by its area. Therefore, we construct the thumbtack which is c times as big by redrawing the original with a "head" that has c times the area. We repeat that the shape of this new area (or of the old one) makes no difference at all.

Analogously, to multiply a sheaf by a scalar c we multiply the density of its lines by that same number. One way of doing this without becoming confused is first to construct a family of planes oriented so as to contain the lines of the sheaf, and spaced by an arbitrary but constant amount; then to rearrange the lines of the sheaf so that each of them lies in one of the planes, and so that the density in all planes is the same. Finally, we multiply the sheaf by c either by increasing the density of planes by this factor (that is, decreasing the distance between them), or else by increasing the density of lines in each plane, leaving the spacing of the planes the same. One must not, of course, do both, since that would multiply the sheaf by c^2 .

To add together two thumbtacks $\underset{\circ}{\boldsymbol{A}}$ and $\underset{\circ}{\boldsymbol{B}}$, we first arbitrarily choose a plane which is *not* parallel to either one, and draw another plane, parallel to the first, and spaced away from it by an arbitrary amount. We also find the orientation of the line that defines the intersection of the two thumbtacks. Using now the fact that we are free to change the shape of a thumbtack, so long

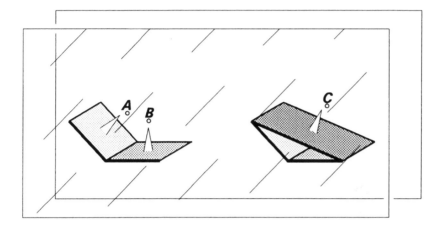

Fig. 3.8: Addition of thumbtacks

as we keep its area constant, we draw both A and B in the shape of parallelograms, two of whose edges are parallel to their intersection, and the other two lie in the previously chosen pair of planes (Fig. 3.8). We also bring the two thumbtacks together so that they have a common edge, taking care that their arrows point in a way that is consistent. The thumbtack C which is the sum of A and B is now obtained by completing the triangle, drawing the new thumbtack also in the shape of a parallelogram between the same pair of planes (Fig. 3.8).

For the case of sheaves, the procedure is a bit more complicated. First, we note that any two sheaves, say $\overrightarrow{A}$ and $\overrightarrow{B}$, determine the orientation of a plane (namely, the plane that contains both the given sheaf directions), and construct a set of such planes, parallel to each other and spaced by an arbitrary but constant spacing (Fig. 3.9). As mentioned earlier, we can now rearrange the two original sheaves so that their lines are all located in those planes, with the same density on each plane.

Concentrating now on one such plane, we have two sets of parallel lines, each with a constant spacing, crossing each other and thus forming a pattern of parallelograms. This immediately determines a new set of parallel lines, namely those which are composed of diagonals of those parallelograms. (In fact, there are two such sets, corresponding to the two diagonals which a parallelogram has; but we choose the lines which run *between* the

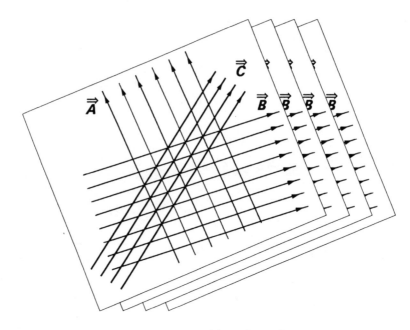

Fig. 3.9: Addition of two sheaves

directions of the original sets when the arrowheads of $\vec{A}$ and $\vec{B}$ are attached.) If these new sets are duplicated on each of the planes, and then viewed as a single sheaf $\vec{C}$, it will, by definition, represent the sum of the original two.

PROBLEMS

3.1 Show that the definition of a cross product of two stacks as given in Fig. 3.7 is not invariant under reflection.

3.2 Develop a rule that assigns an axial sense to the sheaf which is the cross product of two polar stacks. Show that this rule *is* invariant under reflection.

3.3 Is your solution of the previous problem consistent with Fig. 3.7 if the universal right-hand rule is used to convert axial to polar senses? If not, can you change your rule so that it is?

3.4 In Sec. 3.4, we proposed a way of assigning an axial sense to the cross product of two axial arrows. Show that this procedure remains valid when reflected in a mirror.

3.5 If the axial arrow in Fig. 3.4 is converted to a polar one by the universal right-hand rule, do the two arrows have the same senses, or are they opposite? Repeat for the case where it is the polar arrow which is converted to axial.

3.6 Sketch how Fig. 3.4 would look if reflected in a vertical mirror, and repeat the previous problem for that case. Would the situation be different if one used a horizontal mirror instead?

3.7 Is the product of two axial scalars polar or axial? Why?

3.8 In the construction for adding two sheaves (Fig. 3.9), the planes were spaced "by an arbitrary but constant spacing." Show that the result $\overrightarrow{\overrightarrow{C}}$ remains the same when that spacing is changed.

4

COMPLETION OF
THE MENAGERIE

4.1 Reprise: The Need for Invariance

As was already mentioned in Sec. 2.2 in connection with the relation between stacks and arrows, it is always possible — so long as we do not insist on topological invariance — to establish a one-to-one correspondence between any two vector species by equating numerical magnitudes and, if necessary, equating a line-type orientation to a plane-type orientation by the condition of perpendicularity. So we obtain, for example, the arrow that "corresponds" to a given stack by drawing one whose length is equal to the density of sheets in the stack, and whose direction is perpendicular to those sheets. In an analogous manner, we construct an arrow corresponding to a given *thumbtack* by giving it a length equal to the area of the thumbtack and a direction perpendicular to it; or corresponding to a given *sheaf*, by giving

40

the arrow a length equal to the number of sheaf lines per unit area, and a direction parallel to the lines of the sheaf.

The existence of such correspondences naturally makes one wonder whether the long list of sum and product constructions (which is, as we are about to find, even now not yet complete) could not be enormously shortened by transforming all vectors to arrows, using the traditional operations on those arrows, and then converting the results back again. So, for example, if we needed the cross product of two stacks, we would (a) change them into arrows, (b) obtain their product as a traditional arrow cross product (whose length is the area of the subtended parallelogram and so on), and finally (c) convert the product arrow to a sheaf having the direction of the product arrow and a line density equal to its length.

In spite of the indoctrination which this book has been laboring to impose on its readers — that such a procedure is "not admissible" because an arbitrary distortion of space does not preserve perpendicularity or maintain the equality between a length and a density — the curious fact is that the recipe of the previous paragraph will actually work quite well, and produce a result which is topologically invariant, *provided we already know that the cross product of two stacks is a sheaf and not something else.* In that case, but only in that case, the conversions from stack to arrow and from arrow to sheaf cancel each other out in such a way that the final result (but not the intermediate steps) become independent of any space distortion.

To illustrate how this happens, consider the example of two original stacks each having sheets spaced 1 cm apart and perpendicular to each other, and perform a simple scale change from centimeters to meters. In the original system, the two arrows that correspond to the two stacks each have unit length, that is, each is 1 cm long; and their "traditional" cross product also has unit length. But in the modified system, the two stacks each have magnitude 100 (because there are 100 sheets per meter), so that the length of each of the two arrows that correspond to them is 100 m, and their "traditional" cross product is 10,000 m long, which, converted back to centimeters, becomes $1,000,000$ cm — clearly quite a different arrow from what we originally obtained. But if we continue and convert this arrow of length $10,000$ m to a sheaf, we need to draw it so that there are $10,000$ lines per unit area, or $10,000$ lines per m^2. This means that they can be placed on a grid whose spacing is 0.01 m, or 1 cm — exactly the sheaf we would have obtained had we remained with centimeters in the first place.

In other words, the justification for defining the cross product of two stacks as a sheaf lies precisely in the invariance of such a definition to space transformations. Note, however, that even the rather laborious argument of the previous paragraph only proved this invariance for the special case of a uniform change of scale, and considerably more work would be required to generalize it. What's more, we would be forced, in each case, to search the possible vector types for one on which the invariance works, when in fact such a type may not even exist; for example, the cross product of two sheaves simply does not fall into one of the vector categories. That is why we have been so diligent to work only with concepts that can be defined topologically — that is, in terms of unmeasured pictures — in the first place.

4.2 What Is Still Missing?

In Sec. 3.1, we stated that cross products exist only between a pair of arrows or between a pair of stacks, a restriction which remained correct so long as those two categories of vectors were the only two that we knew. But what is really required — as we shall now find — is that the two factors of a cross product have the same *directionality type;* that is, they must both be of the *line type* (either an arrow or a sheaf) or of the *plane type* (either a stack or a thumbtack). Yet that condition alone turns out to be too weak in that, of the six combinations that satisfy it, only four are actually admissible, namely $\vec{A} \times \vec{B}$, $\underline{K} \times \underline{L}$, $\vec{A} \times \vec{\vec{J}}$, and $\underline{K} \times \underset{\circ}{T}$; the remaining two, $\vec{\vec{J}} \times \vec{\vec{Q}}$ and $\underset{\circ}{T} \times \underset{\circ}{S}$, turn out not to work, indicating that there is some other rule which we do not yet understand.

What about dot products? We said originally (Sec. 2.6) that they can be taken only between an arrow and a stack; but again, what is necessary is that the two factors have *opposite* directionality types. This gives four possibilities: $\vec{A}\cdot\underline{K}$, $\vec{A}\cdot\underset{o}{T}$, $\overset{\Rightarrow}{J}\cdot\underline{K}$, and $\overset{\Rightarrow}{J}\cdot\underset{o}{T}$, all of which are possible; but in two of them, the products are not ordinary, but modified, scalars.

From facts such as these it is easy to form the impression that we are pursuing an unending labyrinth, each step forward requiring the introduction of ever new species of quantities, but that is not the case. Indeed, by the end of this chapter it will be apparent that we are actually dealing with a well-defined "menagerie" of exactly seven species, comprising four types of vectors and three types of scalars, which neatly closes upon itself under an equally well-defined set of operations.

4.3 The Remaining Cross Products

First, we need to introduce the remaining two cross products: one between an arrow and a sheaf (resulting in a stack), the other between a thumbtack and a stack (resulting in an arrow).

The construction for the first is as follows (Fig. 4.1). We begin by identifying the orientation of a plane that contains both the given arrow $\vec{A}$ and the given sheaf $\overset{\Rightarrow}{B}$. Within such a plane, we draw lines of the sheaf spaced so that the given arrow takes us from one line to its neighbor. This amounts, of course, to imposing a certain density of those lines within the plane, which we are allowed to do by adjusting the spacing between planes to maintain the given two-dimensional density of the lines, that is, the magnitude of $\overset{\Rightarrow}{B}$. The result of this construction is that each line of the sheaf resides within one of an identical set of planes, spaced by a uniform amount determined by both the given sheaf and the given arrow.

It will be apparent that (a) if $\vec{A}$ is doubled, the spacing of sheaf lines within each plane will also be doubled, which requires that the spacing between the planes be halved; (b) if $\overset{\Rightarrow}{B}$ is doubled (while $\vec{A}$ remains constant), once more the spacing between planes must be halved to accommodate all the lines of the sheaf. As for the directional relationship, it is also clear that (c) if the direction of $\vec{A}$ is brought closer and closer to that of $\vec{\vec{B}}$ (while both magnitudes are kept constant), the actual spacing of the lines within the plane will get smaller and smaller; hence, the density of the planes must also become smaller and smaller. In view of

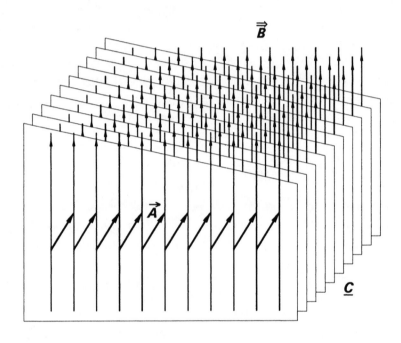

Fig. 4.1: Cross product of an arrow and a sheaf

this behavior, it is not surprising that the *the stack $\underline{C}$ defined by this system of planes can be taken to be the cross product of the arrow $\vec{A}$ and the sheaf $\vec{\vec{B}}$*. Its sense is either polar, determined by the universal right-hand rule (Fig. 3.6), or axial, indicated by an appropriate curlicue.

For the second construction − the cross product between a thumbtack $\underset{\circ}{A}$ and a stack $\underline{B}$ − we place the thumbtack between two neighboring sheets of the stack and deform it into a parallelogram a pair of whose opposite sides lie in those two sheets (Fig. 4.2); we know, of course, that this can be done because it is only the *area* of the thumbtack that is fixed. The arrow $\vec{C}$ formed by the line segment of one of those parallelogram sides, namely a side which lies in one of the stack sheets, is then the cross product.

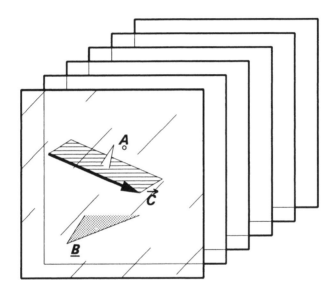

Fig. 4.2: Cross product of thumbtack and stack

In Fig. 4.2, we have shown the sense of $\vec{C}$ by an arrowhead; that is, we have made it polar. The (perhaps more correct) *axial* sense would be indicated by a curlicue in the direction from A to B, but we have converted it by the universal right-hand rule.

4.4 More Dot Products: Scalar Densities and Capacities

With regard to dot products, we have until now defined only one (the one between an arrow and a stack), but as more vector types have since appeared on the scene, the question needs to be reopened. In fact, three new products are now possible, of which the simplest is between a sheaf and a thumbtack; it is simply *the number of lines of the sheaf that thread the thumbtack* (Fig. 4.3). Like the one between an arrow and a stack, the result here is obtained by a simple operation of *counting,* and is therefore invariant to all continuous space transformations (since no such transformation can convert a line that does not thread a given closed curve into one that does). In Fig. 4.3, $\vec{A} \cdot \vec{B}$ is equal to 9.

The sign of this new dot product is positive if the lines of the sheaf pass through the thumbtack in the thumbtack's own sense,

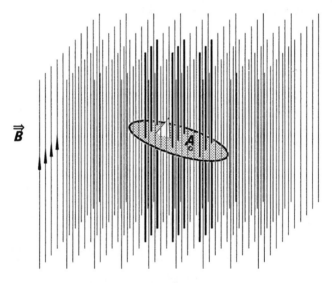

Fig. 4.3: Dot product of sheaf and thumbtack

and negative otherwise. It goes without saying that in this, as in all similar cases, when one (but not both) of the factors is axial, the result will be an axial scalar, or pseudoscalar.

Consider now the superposition of a sheaf and a stack, whose intersections define a "dust cloud" of a certain density; it is this density that we shall take to be their dot product. It is not a simple scalar, however, since it is determined, not by merely counting objects (such as lines threading a loop, or sheets spanned by an arrow) but counting objects *within a unit volume*. Accordingly, while not invariant to arbitrary transformations of space, its variation is particularly simple, depending only upon the size of the volume element. For our purposes we could invent a picturesque name for this new type of quantity, such as, for example, a *swarm;* but, for reasons which shall soon become clear, we now forgo picturesque language in favor of the more prosaic term *scalar density.*

There remains one possible dot product to consider, formed between a thumbtack and an arrow (in this, as in all other dot products, the order of the two factors is immaterial). It is defined as *the volume swept out by the given thumbtack when it is displaced by the given arrow* (Fig. 4.4), and typifies a quantity

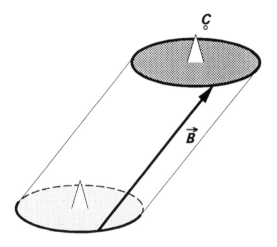

Fig. 4.4: Dot product of thumbtack and arrow

called a *scalar capacity.* Like a scalar density, it is not invariant to space transformations, but changes according to the size of the volume element; but since a scalar density is a number of objects *per volume,* and a scalar capacity is itself *a volume,* the two vary in ways that are exactly reciprocal to each other. In fact, we may note that *the product of a scalar density and a scalar capacity is a simple scalar.*

4.5 Formalization of Nomenclature

We have now completed the introduction of our *sevenfold menagerie* of quantities, comprising four kinds of vectors and three kinds of scalars (the number would, of course, be doubled if we listed polar and axial species separately). In many cases, we have given them pictorial nicknames (such as "arrow," "stack," and so on). The time has now come to replace those nicknames with the more formal names in general use, as are shown in the following table:

PICTORIAL	FORMAL
arrow	contravariant vector
stack	covariant vector
sheaf	contravariant vector density
thumbtack	covariant vector capacity
?	scalar
swarm?	scalar density
?	scalar capacity

This new terminology appears to imply a number of relationships that we have not previously noted. We first list these as purely verbal inferences, then proceed to justify them.

●   Vectors are classified as "contravariant" if their directionality is of the *line type,* and "covariant" if it is of the *plane type;* the choice of prefixes implies that those directionality types are in some sense opposites of each other.
●   The oppositeness of the directionality types is supported by the observation that a dot product, which combines one of each type, yields a product which is directionless — that is, one of the scalars.

• The fact that (as we saw) the product of a scalar *density* and a scalar *capacity* is a *plain* scalar suggests that "density" and "capacity" function as opposites in another, independent dimension, one which works for dot products as well. Thus, for example, the dot product of a contravariant vector *density* (sheaf) and a *plain* covariant vector (stack) is a scalar *density*, and that of a contravariant vector *density* (sheaf) and a covariant vector *capacity* (thumbtack) is a *plain* scalar. The same rule works for the other cases.

• The situation for cross products is rather more complicated, in that it is necessary to attribute nonzero values, both on the density and the variance dimensions, to the symbol $\times$ itself; specifically, it must be considered as *either* a triple-contra density, *or* a triple-co capacity. For example, in the equation $\vec{A} \times \vec{B} = \underset{\circ}{C}$ the second choice has been taken, so that the two "contras" and the triple "co" combine to a single "co," and the capacity remains. The result is, of course, a covariant vector capacity, or a thumbtack. The reader can verify this rule for the other cross products also, and note that the combinations which we said were impossible, $\vec{\vec{J}} \times \vec{\vec{Q}}$ and $\underset{\circ}{T} \times \underset{\circ}{S}$, would require a covariant vector density or contravariant vector capacity for their result, neither of which exists in our menagerie (see Sec. 4.6).

• The preceding rule may strike one as so complex as to make it preferable to memorize the result and discard the rule, which may well be so. We have, nonetheless, stated it for the benefit of those who may, at a later stage, move into the more general lore of spaces of N dimensions, when the corresponding extended rule becomes useful.

4.6 The Geometrical Meaning of "Co-" and "Contra-"

The observation that "co-" and "contra-" refer to the two possible vector directionality types is correct, but that is not where the terminology originated; rather, the prefixes describe what happens to the magnitude of a vector when its space is squeezed. If the compression takes place along the direction of the vector, it causes "co-" vectors to become larger in proportion to the amount of squeezing (because their sheets are brought closer together), whereas "contra-" vectors becomes smaller in the same proportion (because their arrows are shortened). This explains, of course, why the dot product of one contravariant and one covariant vector is a scalar, that is, a quantity which does not change at all when its space is compressed.

The situation is modified, however, if the word "density" (or its negative, "capacity") appears in the description of the vector. To understand this modification, consider first the scalar density and the scalar capacity. Neither one of them has any direction, of course; but they *are* modified when space is squeezed along some direction, the density increasing and the capacity decreasing, both in proportion to the amount of linear compression.

Given this behavior, it is not surprising that a "contravariant vector density" is one which combines the properties of a contravariant vector and a scalar density: if its space is squeezed along the direction of the vector, its magnitude is decreased in proportion (because it is contravariant) but also increased in the same proportion (because it is a density); consequently, it does not change at all. But if we do our compression in a transverse direction, the magnitude does not change (because it is a vector) but it also increases in proportion (because it is a density); with the result that it increases in proportion. It is easy to recognize that this is exactly the behavior of a sheaf; and a moment's thought will reveal that the same logic also works, with opposite result, for a covariant vector capacity (thumbtack).

We can now finally understand why our menagerie is limited to the four vector types; why we do not, for example, include such a thing as a "covariant vector density." It is simply that such an entity would, when squeezed either along its direction or perpendicularly to it, need to vary nonlinearly with the amount of compression, making it impossible to represent it by a simple picture.

4.7 Multiplication by Scalar Densities and Capacities

There remain a few multiplication operations, those between a scalar and a vector where densities are involved, which we have not explicitly discussed. Although the present section will, for the sake of completeness, describe the necessary operations, the reader should understand that there is no good reason for memorizing them: *Provided that a certain product exists,* it is straightforward to search out the corresponding construction simply by looking for one that will produce the required type of object. So, for example, if we are asked to construct the product of a scalar density and a contravariant vector density, we should know immediately that, since there is no such thing as a "contravariant double-weight density vector" in our menagerie, the requested product does not exist. (The reason for the "non-existence" of double-weight densities is, of course, the same as was discussed in the previous section.)

The constructions for the products that do exist are as follows:

• Arrow times scalar density: organize the "dust particles" of the scalar density along lines such that within each line they are spaced by the given arrow. The result is a bundle of lines, which comprises the required sheaf.
• Sheaf times scalar capacity: organize the lines of the sheaf so that, on an arbitrary plane that cuts them, the intersections form a lattice of parallelogram-shaped cells. Make the given scalar capacity (which is a volume) into a parallelepiped whose base is one of those cells, and whose remaining edge follows the direction of a sheaf line. That remaining edge is the required product, which is an arrow.
• Stack times scalar capacity: form the capacity (which is a volume) into a cylinder which just fits between two sheets of the given stack. The base that lies in the stack sheet is the required thumbtack.
• Thumbtack times scalar density: form the thumbtack into a

parallelogram, and construct a plane filled with such parallelograms. Place one dust particle in the center of each, and space such identical planes so that the density of dust particles is the given scalar density. This collection of planes is the required stack.

One should also remark that, because scalars have well-defined reciprocals, it is possible not only to multiply but to divide by one of them. In this way one obtains four more constructions which are, however, each equivalent to one of those given above.

PROBLEMS

4.1 Beginning with the picture of a scalar density ρ as a "dust cloud," and of a scalar capacity V as a volume, find a purely geometrical definition for the product ρV. It must, of course, be a pure scalar, independent of distortions of space.

4.2 Beginning with the construction of Fig. 4.2, show by a purely geometrical construction that the cross product $(\underset{\circ}{A}+\underset{\circ}{B})\times\underline{C}$ is the sum of $\underset{\circ}{A}\times\underline{C}$ and $\underset{\circ}{B}\times\underline{C}$.

4.3 Give an analogous proof for $(\underset{\circ}{A}+\underset{\circ}{B})\cdot\vec{C}$.

4.4 Consider the "triple scalar product" $\vec{A}\cdot\vec{B}\times\vec{C}$ of three arrows. Explain why its meaning is unambiguous in spite of the absence of parentheses to specify the order of multiplication. To what species of the menagerie does this triple scalar product belong?

4.5 Prove (geometrically, of course) that $\vec{A}\cdot\vec{B}\times\vec{C}=\vec{A}\times\vec{B}\cdot\vec{C}$.

4.6 Discuss the triple scalar product $\underline{A}\cdot\underline{B}\times\underline{C}$ of three stacks. Prove that $\underline{A}\cdot\underline{B}\times\underline{C}=\underline{A}\times\underline{B}\cdot\underline{C}$.

4.7 Is the triple scalar product $\vec{A}\cdot\vec{B}\times\vec{C}$ definable in our terms? How about $\vec{A}\times\vec{B}\cdot\vec{C}$?

4.8 Let ρ be a scalar density ("swarm"). Given two arrows $\vec{A}$ and $\vec{B}$, show that $(\rho\vec{A})\times\vec{B}=\vec{A}\times(\rho\vec{B})$.

4.9 Let σ be a scalar capacity. Given a stack $\underline{A}$ and an arrow $\vec{B}$, show that $(\sigma\underline{A})\cdot\vec{B}=\sigma(\underline{A}\cdot\vec{B})$.

5

FIELDS AND THE
GEOMETRICAL CALCULUS

5.1 Fields

A *field* is a region of space for each point of which a value of a
certain quantity is defined; we can speak, for example, of an
arrow field, or a thumbtack field, or a scalar field. (In formal
terminology the first two are, as we now know, a contravariant
vector field and a covariant vector capacity field.) A field can be
represented pictorially by sprinkling space with geometrical
symbols of the corresponding quantity; for example, we picture
an arrow field as a "forest" of arrows covering space, the length
and direction of each one corresponding to the value of the arrow
field at that location. In doing so, however, we must again face
the question of the *size of symbols*. Specifically, what are we to
do if the value of an arrow field is (appreciably) different at the
locations of the two ends of one of its arrows? Or if a stack field
is so weak that its value is (appreciably) different from one sheet
to the next?

The answer to this problem is essentially the same as we have already discussed in Sec. 2.3 in connection with space transformations: it is always possible to choose a scale which makes the symbol as small as we wish. For example, we can always interleave a stack with nine more sheets for each that already exists, increasing their density by a factor of ten; at the same time, we change the scale by which stacks are measured so that the new one represents the same physical quantity as the old one did. Since it only takes two sheets to define a stack, its picture can now be ten times as small. In this way we can continue subdividing until our criterion is satisfied – namely, that a number of sheets of the stack can fit into a region of space across which the value of the field does not appreciably vary.

As a second example, a scalar density field can be represented by a "dusty" region of space; again, we assume (as we have a right to do) that many "dust particles" appear within a volume across which the value of the field is pretty much constant. Incidentally, our previous requirement that space transformations be continuous and differentiable now becomes a restriction on the field itself.

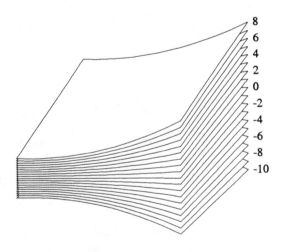

Fig. 5.1: A scalar field

A *scalar,* being simply a number, has not up to this point required any geometrical representation, but for a scalar *field* it is convenient to have one. We shall use a set of "equipotential surfaces," that is, surfaces on which the scalar in question is constant, and between which it changes by a fixed increment (Fig. 5.1). Again, this increment is assumed to have been chosen sufficiently small so that the surfaces are nearly flat, and are spaced by a nearly constant amount, within a region that contains many of them.

5.2 The Gradient

Every scalar field also automatically represents a stack field. To understand this somewhat startling fact, let a scalar field Φ be

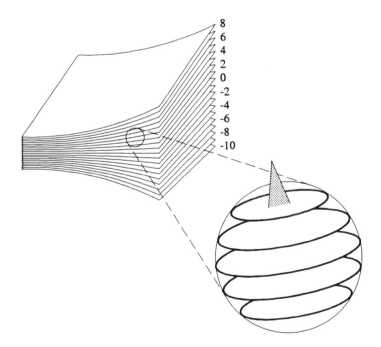

Fig. 5.2: Construction of gradient

represented by a family of equipotential surfaces (we assume, as in the previous section, that their scale is chosen so that their spacing and orientation are constant over the space of many sheets). Imagine now that we take a tool such as a "melon baller" (used to cut small spherical chunks out of a melon) and cut out, or delineate, a small region of our space, which will now contain a number of pieces of surface approximately parallel to each other and approximately evenly spaced (Fig. 5.2); right then and there we have a stack $\underline{k}$ associated with that location. As for the sense of the stack, its arrowhead is taken to point from smaller to larger values of the original scalar field. If Φ is an axial scalar, then $\underline{k}$ is, of course, an axial stack; we draw its curlicue so as to be clockwise when viewed in the direction in which Φ is becoming more right-handed (or less left-handed).

We describe the connection between $\underline{k}$ and Φ by saying that $\underline{k}$ is the *gradient* of Φ, and symbolize the relation in equation form as

$$\underline{k} = \text{grad} \, \Phi \, . \tag{5.2.1}$$

A simple integral identity follows from this definition, namely

$$\Phi_2 - \Phi_1 = \int_1^2 (\text{grad} \, \Phi) \cdot d\vec{r} \, , \tag{5.2.2}$$

where Φ_1 and Φ_2 are the values of Φ at the two points 1 and 2, the integral is a line integral along any path which goes from 1 to 2, and $d\vec{r}$ is an infinitesimal arrow connecting two neighboring points on the path. To see why this equation is true, recall that $\underline{k} \cdot d\vec{r}$ means, by definition, the number of sheets of the stack $\underline{k}$ spanned by the arrow $d\vec{r}$; and since the sheets of this stack are the same as the equipotential sheets of Φ, $\underline{k} \cdot d\vec{r}$ becomes the number of equipotential sheets spanned in the infinitesimal displacement $d\vec{r}$, that is, the change in the value of Φ corresponding to the displacement. Hence, in integrating this differential from point 1 to point 2 we obtain the *total* change in the scalar Φ.

Although the integral identity (5.2.2) is often described as a "theorem," when formulated geometrically its logic is so obvious that it hardly merits that name. A simple corollary is that, since the left side of Eq. (5.2.2) does not depend on the path chosen to connect the two points, neither does the right side. In particular, if we perform the line integration around a closed path, the result must always be zero.

5.3 The Curl

The construction of the gradient in the previous section naturally leads us to ask the converse question: Suppose we are given a stack field which varies continuously in space. Is it possible to

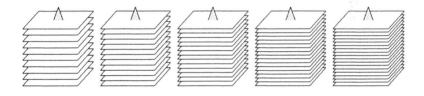

Fig. 5.3: A stack field which cannot continuously join

arrange its "microsheets" so as to join into macroscopic sheets which could then be interpreted as the equipotentials of a scalar? In other words, can an arbitrary stack field be viewed as the gradient of a scalar? The answer, in general, is "No." To make this clear, consider the following counterexample (Fig. 5.3): Let a stack field have a constant direction, so that its sheets are, say, always horizontal; and let its magnitude increase along a horizontal direction, say from left to right. Obviously, the only way in which these individual stack sheets could seamlessly join together would be to bend away, from time to time, from the horizontal direction, but then the field would no longer have the constant direction which was specified for it.

It is always possible, however, simply to join the "microsheets" into large surfaces *as much as possible;* which means that, from time to time, a new surface will need to originate (Fig. 5.4), generating a "loose edge." It is clear that these loose edges will themselves constitute a sheaf field, whose value at each point in space can be obtained by applying our "melon baller" to the pattern of those edges. The sense of this sheaf field is, properly speaking, axial, where we can (for

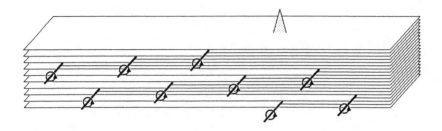

Fig. 5.4: Definition of the curl

example) draw our curlicue for each line by looking at the stack sheet which ends there, beginning on the side of the sheet from which its own arrow originally emerged, and curling around to the other side, as in Fig. 5.4. (Naturally, if the stack field was axial, then the new field is polar.)

The sheaf field obtained by this construction is called the *curl* of the original stack field; if we name the new field $\vec{\vec{S}}$, and the original one $\underline{k}$, then the relation is written in equation form as

$$\vec{\vec{S}} = \text{curl}\,\underline{k}\,.\qquad(5.3.1)$$

An obvious identity follows from the realization that if $\underline{k}$ was itself the gradient of some scalar field Φ, then its sheets can be put back together without any loose edges arising, so that

$$\text{curl}\,\text{grad}\,\Phi = 0\,,\qquad(5.3.2)$$

where Φ is any scalar. Stated in words, *the curl of any gradient is identically zero.*

A more intricate extension of the same relation is obtained if we consider the line integral along a closed loop

$$\oint \underline{k} \cdot \vec{dr},$$ (5.3.3)

which we previously showed to vanish if $\underline{k}$ is a gradient. If it is not, the differential $\underline{k} \cdot \vec{dr}$ is still a count of the sheets of $\underline{k}$ spanned by the differential displacement $\vec{dr}$. Now if we integrate this around a closed loop, the only way in which the number of sheets crossed going in one direction can fail to be canceled by the number that are crossed on the way back is for some of those sheets to terminate inside the loop (Fig. 5.5). To count them, we draw an arbitrary surface spanning the loop, divide it into infinitesimal thumbtacks $d\underset{\circ}{\Sigma}$, and sum the number of loose edges

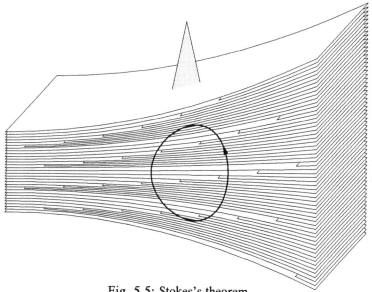

Fig. 5.5: Stokes's theorem

threading each of the thumbtacks. Recalling the definition of a dot product of a sheaf and a thumbtack, we thus obtain the identity

$$\oint \underline{k} \cdot d\vec{r} = \iint (\text{curl } \underline{k}) \cdot d\underline{\Sigma} \, , \tag{5.3.4}$$

where the line integral on the left is taken around an arbitrary closed loop, and the surface integral on the right over any surface which spans that loop. It follows as a corollary that since the first is independent of the surface chosen to span the given loop, the same must be true of the second. For the particular case of Fig. 5.5, the loop crosses 35 sheets on the way up (that is, in the direction of the stack) and 30 on the way down, resulting in a value of $+5$ for either integral.

In traditional treatments, Eq. (5.3.4) is generally called "Stokes's theorem"; once more, however, its geometrical logic is so straightforward as hardly to merit the name.

5.4 The Divergence

The corollary just mentioned can be rewritten as

$$\iint_S (\text{curl } \underline{k}) \cdot d\underline{\Sigma} = 0 \, , \tag{5.4.1}$$

where S now represents a *closed* surface, and the direction of the differential thumbtack $d\underline{\Sigma}$ is consistently *outward* from the volume enclosed by the surface. Equation (5.4.1) thus states that no net number of lines of the sheaf field curl $\underline{k}$ emerge from any closed surface. Geometrically, this is obvious, since the lines of curl $\underline{k}$ are the "loose edges" of the sheets obtained by joining together (to the extent possible) the "microsheets" of the stack field $\underline{k}$, and there is no way in which an edge of a surface can end; either it closes on itself, or it runs off the picture. Therefore any such loose edges that emerge from the closed surface must also have entered it.

On the other hand, the statement that curl $\underline{k}$ is constructed from continuous "streamlines" which never begin or end cannot be made about an arbitrary sheaf field $\vec{J}$. In fact, we can ask the question analogous to the one of the previous section: given an arbitrary (but continuous) sheaf field, how well can we do in connecting its "microlines" into continuous macroscopic "streamlines"? Clearly, such a construction will, from time to time, have streamlines that begin and end; equally clearly, the density of these loose ends will itself be represented by a scalar

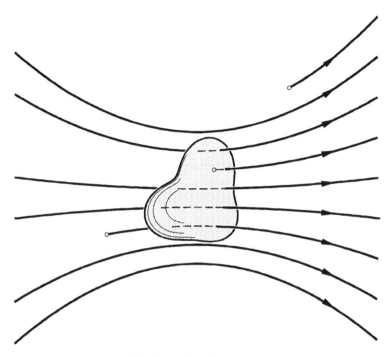

Fig. 5.6: Gauss's theorem

density, which we define as positive where the streamlines of $\vec{\vec{J}}$ originate, and negative where they terminate. This scalar density ρ, which can be computed for any given sheaf field, is called its *divergence;* in symbols,

$$\rho = \operatorname{div}\vec{\vec{J}}. \tag{5.4.2}$$

If $\vec{\vec{J}}$ was itself the curl of some stack field $\underline{k}$, then it has no divergence, so that we immediately have

$$\operatorname{div}\operatorname{curl}\underline{k} = 0 \tag{5.4.3}$$

for any covariant vector field $\underline{k}$.

As in the cases of the gradient and the curl, the definition of the divergence immediately leads to an integral identity, as follows. If we consider the integral over a closed surface of the dot product $\vec{J} \cdot d\Sigma$, where the differential thumbtack $d\Sigma$ is the element of surface (with its sense, as always, pointing outward), what we obtain is the net number of streamlines of $\vec{J}$ which emerge from the surface. This number can differ from zero only if there is a net number of such lines of $\vec{J}$ which are born (or die) inside the surface (Fig 5.6). But since $\mathrm{div}\,\vec{J}$ is the density of such beginnings or endings, we can obtain the same net number by multiplying $\mathrm{div}\,\vec{J}$ by the differential scalar capacity $d\tau$ that represents an element of volume and integrating over the volume enclosed by the surface; in other words

$$\iint \vec{J} \cdot d\Sigma = \iiint (\mathrm{div}\,\vec{J})\, d\tau \qquad (5.4.4)$$

for any closed surface. This relation is sometimes known as *Gauss's theorem*.

5.5 The Inverse Operations

The integral identity (5.2.2) can be used to formulate an operation inverse to the gradient, that is, an operation to find the function $\Phi(\vec{r})$ of position when its gradient $\mathrm{grad}\,\Phi$ is given. We simply write it as an indefinite line integral:

$$\Phi(\vec{r}) = \int (\mathrm{grad}\,\Phi) \cdot d\vec{r}. \qquad (5.5.1)$$

Of course, the given stack field must have a curl which vanishes everywhere; were this not so, the line integral would not be independent of path, making Eq. (5.5.1) meaningless.

Since an indefinite integral always has an arbitrary constant added to it, the inverse of a gradient can be determined only to within an additive constant. Indeed, this is obvious from the gradient's definition, because a constant added to Φ only changes the *labeling* of its equipotential surfaces, but not their geometrical pattern. Alternatively, the constant can be made explicit by changing Eq. (5.5.1) to a *definite* line integral which runs from an arbitrarily chosen fixed point $\vec{r}_0$ to the variable point $\vec{r}$.

The corresponding operation for the curl — that is, the operation which constructs a stack field given the sheaf field that is equal to its curl — is conveniently defined geometrically by the "hanging laundry" construction. To carry it out, we note first that the given sheaf field must be divergenceless, that is, must be comprised of continuous, endless streamlines, or it could not possibly be the curl of anything at all. We then visualize these given streamlines as rigid laundry lines, and hang a semi-infinite sheet from each of them, extending straight down under an imagined force of gravity. These sheets then become the sheets of the desired stack field, since by definition the curl is made up of the loose edges of the stack field whose curl is to be taken.

The "hanging laundry" construction produces a stack field whose sheets are everywhere parallel to an arbitrarily chosen vertical direction, showing us immediately that the solution to the inverse curl problem is by no means unique. In fact, we may add to it any stack field whatsoever whose curl is zero, that is, any stack field which is itself the gradient of something.

Finally, we examine the inverse divergence problem. In this case, we are given a scalar density — a dust cloud, or "swarm" as we called it earlier — consisting of some "particles" which are negative, and some which are positive. To find a sheaf field whose divergence is given by this "swarm," we need merely to connect each positive particle to a negative particle by an arbitrary continuous curve, and consider the resulting family of curves as the streamlines of the desired sheaf field. As in the case of the curl, the non-uniqueness of this solution is seen immediately to be immense. In fact, any one of those solutions can be changed into any other by adding another sheaf field which has zero divergence, that is, whose streamlines never begin or end.

5.6 The Meaning of Differential Operations

The reader will be aware, of course, that in traditional treatments the gradient, curl, and divergence are considered *differential* operations in space, as originally specified by Eqs. (1.2.4-6); in other words, they measure, in one way or another, the extent to which a given field differs between neighboring points. In particular, we know that all three are identically zero if the field to which they are applied is constant in space (the converse need not, of course, be true). Although our purely geometrical treatment has not emphasized the kind of algebraic operations that are usually associated with differentiation, it is easy to verify that the correspondence nonetheless holds.

Consider, first, the gradient: it always operates on a scalar field, represented by a family of equipotentials. As such a scalar field becomes more and more spatially constant, its equipotentials move further and further apart. Applying our "melon baller" to such a weakly varying field will produce a stack field whose sheets are correspondingly far apart, that is, one which is itself small, tending closer and closer to zero as the spatial variation of the original field gets weaker and weaker.

For the curl (divergence), it is similarly obvious that if a stack (sheaf) field is constant in space, its "microsheets" ("microlines") will join perfectly into macroscopic sheets (lines) without loose edges (ends). In either case, then, the curl (divergence) approaches zero as the field to which the operator is applied approaches constancy.

At the same time, such a discussion makes clear that trying to describe everything in pure pictures, though providing immense insight into the nature of the relationships, may not be suitable for more precise computations, ultimately depending (as it does) on the absolute accuracy with which we are able to draw. We therefore go on, in the next chapter, to develop concepts that will make it possible finally to apply the power of numerical computation to the geometrical quantities with which we have now become familiar.

PROBLEMS

5.1   Construct an explicit geometrical recipe for finding the (axial) sense of the curl of a polar stack field.

5.2   Construct an explicit geometrical recipe for finding the (polar) sense of the curl of an axial stack field.

5.3   Restate your answers to the previous two problems in terms of a right-hand rule, for people who don't like to use axial senses (but, of course, don't mind abandoning invariance under reflection).

5.4   One can define the total electric charge within a region as "the number of electric field lines which leave the region," and the electric charge density as "the density with which electric field lines originate." What flavor do these definitions imply for the electric field?

5.5   One of the fundamental laws of electrostatics is that the electric field E is *conservative;* that is, it has zero curl. What flavor does this imply for the electric field? Is it the same as in the previous problem? Compare also Problems 1.1-3.

6

COORDINATES
AND COMPONENTS

6.1 Coordinate Systems

Since any two scalar fields, say q_1 and q_2, are each associated with a family of equipotential surfaces, the intersections of those surfaces — that is, the loci of points where the values of both q_1 and q_2 are specified — will be curves. If we add a third scalar field q_3, the intersections of the surfaces of all three form a family of points. Thus any point in space can generally be specified by giving the value of q_1 for the q_1-surface that it is on, the value of q_2 for the q_2-surface that it is on, and the value of q_3 for the q_3-surface that it is on; in other words, the set of three quantities $\{q_1, q_2, q_3\}$ comprise a *coordinate system.*

The three families of surfaces corresponding to such a system will divide space into a family of *cells,* as in Fig. 6.1. If the scale is chosen to be sufficiently fine — in other words, if the interval Δq that separates each surface from its neighbors is sufficiently small — then these cells approach parallelepipeds which are more or less identical to each other throughout a region containing many of them (see the discussion of Sec. 5.1). Such a construction gives rise, at every point in space, to a natural *basis*

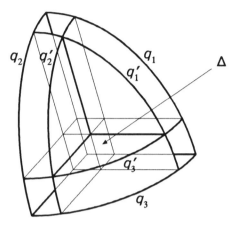

Fig. 6.1: Formation of the unit cell

for each of the types of vector and scalar quantities that we have been considering, as follows (see Fig. 6.2):

• The three cell edges issuing from a point define three linearly independent arrows;
• The three pairs of opposing cell faces define three linearly independent stacks;
• The three cell faces meeting at a point define three linearly independent thumbtacks;
• The three sets of parallel cell edges, four edges to each set, define three linearly independent sheaves;
• The eight corners define a scalar density;
• The volume of the cell defines a scalar capacity;
• Last but by no means least, pure scalars have absolute numerical values and so do not require any "natural basis."

Some precautions do need to be taken. We must not allow

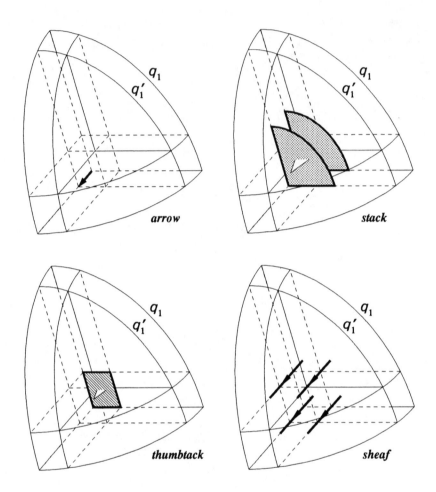

Fig. 6.2: How the four vector bases arise

two of the coordinate surfaces to coincide or even be tangent to each other, since in the immediate vicinity of the point of tangency the volume of the unit cell would vanish and information would be lost. Nor can we allow the type of indeterminacy that results from an equipotential surface crossing itself. Finally, we must avoid pathologies such as an equipotential surface having a sharp "crease," or otherwise giving rise to a discontinuity of the coordinate system. It should be understood that in many important physical applications pathologies *do* exist but they are localized; for example, in the cylindrical coordinate system (Sec. 6.6), all surfaces of constant ϕ intersect on the ζ-axis. This can by no means be ignored, and all sorts of non-analyticities do occur at that location, but we need not worry about them now.

Once a basis has been established for a particular type of quantity, the value of such a quantity can be specified numerically by giving the coefficients that define it as a linear combination of the basis members; these coefficients are known as the *components* of the quantity. (In the case of scalar densities or scalar capacities, the basis has only one member, and so the quantity has only one "component.")

6.2 The Bases of Scalar Capacities and Densities

There is a simple numerical relationship between the bases for scalar capacities and densities as we have just defined them. This is most easily seen by imagining the "dust particles" which are located on all the unit cell corners displaced slightly along a body diagonal of the cell, making it clear that there is one and only one such particle for each cell. Thus the product of the density of particles and the volume of the cell is exactly one, so that the two bases are reciprocals of each other. We shall denote the volume of a unit cell by the symbol Δ, and the density of corners by $1/\Delta$. In general Δ is, of course, a function of the three coordinates $\{q_1, q_2, q_3\}$; this introduces no ambiguity, however, so long as our interest is in field quantities, which are themselves defined as functions of location. It is then taken for granted that the bases to be used are those that correspond to that same point in space.

The purpose of defining bases is, as we said, that it makes possible the assignment of numerical values to densities or capacities while still maintaining our "topological" framework, that is, while continuing to describe those objects without resorting to rulers or protractors; instead, we specify the magnitude of a scalar capacity (or density) by specifying the component, that is, the numerical factor that relates it to its basis. Such an approach does, however, demand some care, because the behavior of the resulting quantities is different depending on which of the two choices of Sec. 1.6 we take; that is, whether "transformation properties" are determined by the way quantities behave under coordinate transformations, or by the way they behave under distortions of the system.

Taking the second possibility first, let's imagine some apparatus containing a number of separated chambers, and suppose that we are interested in specifying the volumes of those chambers numerically, which ordinarily cannot be done without using rulers and protractors. (Strictly speaking, protractors are never more than a convenience, since the angles of a triangle can also be obtained from a knowledge of the lengths of its sides.) But if we know that a certain coordinate system is "embedded" in the apparatus, and we specify each volume by giving the number of unit cells that it contains, then that number will not change if the system is distorted, since the coordinate system will be distorted in the same way as the apparatus. According to such an interpretation, the numerical specification of the volume of a chamber, which we called its component, becomes a scalar — that is, its value is independent of distortions.

By contrast, if our coordinate system is not "embedded" in the apparatus but is itself subject to change while the apparatus remains the same — for example, if we first use a Cartesian system, then change over to a spherical one — then the numerical relation of the volume of one of our chambers to the volume of the unit cell *does* change. With such a picture, the numerical volume of a chamber is *not* a scalar. Yet its variation is extremely systematic, in that *different* scalar capacities all vary in *the same* way, entirely determined by the very statement that *so-and-so is a scalar capacity*. It is this latter interpretation, in which the nature of numerical quantities is essentially defined by the way they change under a change of coordinate system, that we shall always use from now on.

6.3 The Arrow and Stack Bases

In Sec. 6.1, we indicated the possibility of defining an arrow basis directly from a coordinate system by using the fact that the three cell edges issuing from any point define three linearly independent arrows. To make this definition precise, we note, as in the left portion of Fig. 6.3, that the first of the cell edges lies along the intersection of the two equipotential surfaces

$$q_2 = \text{const} , \qquad q_3 = \text{const}' ; \qquad\qquad (6.3.1)$$

if we assume that neighboring equipotentials of q_1 are spaced by unity, then the length of this cell edge is determined by

$$\text{const}'' < q_1 < \text{const}'' + 1 . \qquad\qquad (6.3.2)$$

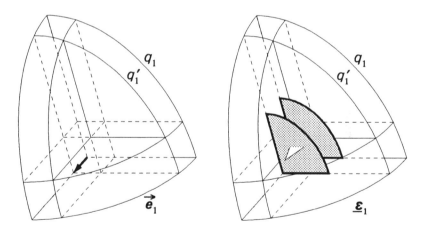

Fig. 6.3: Arrow basis and stack basis

This definition assumes, of course, that the scale chosen for the qs is fine enough.

Although convenient for visualization, such a restriction is not actually required, and this is probably a good time to free ourselves of it. The problem lies in the number "1" in Eq. (6.3.2), in that we need to assume that in this interval all the coordinate surfaces remain essentially flat and equally spaced; but we can do just as well by substituting an increment Δq that is infinitesimal, leading to a cell edge that is also infinitesimal, and then enlarging the result by the reciprocal factor $1/\Delta q$. The reader will, of course, recognize this as nothing other than the familiar process of differentiation; in other words, without assuming anything about the rate at which the coordinate surfaces vary (except that they do so in a continuous and differentiable manner), we directly *define* the three arrow basis vectors $\vec{e}_i$ by the formulas

$$\vec{e}_i \equiv \partial\vec{r}/\partial q_i, \qquad i = 1, 2, 3 .\tag{6.3.3}$$

Here the "radius vector" $\vec{r}$ is an arrow whose tail is at some fixed point, and whose head is at the (variable) point in question; and the partial derivative indicates, as usual, that one q is varied while the other two remain constant — that is, that the head of $\vec{r}$ moves along the intersection of two equipotential surfaces.

Returning now to the fundamental pictorial definitions of Sec. 6.1, it becomes clear that the three vectors of the stack basis, which should correspond to the stacks defined by pairs of opposite faces of our parallelepipedal unit cell, can also be defined in terms of already familiar operators, namely the operator that we called the *gradient* (as in the right portion of Fig. 6.3). Specifically, those three basis vectors, which we shall call $\underline{\varepsilon}_i$, are nothing but the gradients of the respective coordinates, that is,

$$\underline{\varepsilon}_i \equiv \operatorname{grad} q_i, \qquad i = 1, 2, 3 .\tag{6.3.4}$$

There is a basic identity relating the $\vec{e}$s and the $\underline{\varepsilon}$s. Since $\vec{e}_1$ lies along the intersection of a $\underline{\varepsilon}_2$-sheet and a $\underline{\varepsilon}_3$-sheet, the two dot products $\vec{e}_1 \cdot \underline{\varepsilon}_2$ and $\vec{e}_1 \cdot \underline{\varepsilon}_3$ both vanish. On the other hand, the dot product $\vec{e}_1 \cdot \underline{\varepsilon}_1$ is the number of sheets of $\underline{\varepsilon}_1$ (that is, of q_1)

spanned by the arrow $\vec{e}_1$, which is, of course, exactly 1. Generalizing these relations to the other members of the two bases, we obtain the *orthonormality relations*

$$\vec{e}_i \cdot \underline{\varepsilon}_j = \delta_{ij} \ . \tag{6.3.5}$$

The symbol on the right is called the "Kronecker delta"; by definition, it is equal to 1 if its two indices are equal, and to zero otherwise.

6.4 Stack-Arrow Dot Product in Terms of Components

As an illustration of the use of components, consider the dot product of an arrow and a stack. Suppose that each of these is specified in terms of its components with regard to the appropriate basis:

$$\vec{A} = A_1\vec{e}_1 + A_2\vec{e}_2 + A_3\vec{e}_3 \ , \tag{6.4.1}$$
$$\underline{B} = B_1\underline{\varepsilon}_1 + B_2\underline{\varepsilon}_2 + B_3\underline{\varepsilon}_3 \ . \tag{6.4.2}$$

The dot product of the two, if multiplied out by the distributive law, looks as though it might contain nine terms; but because of the orthonormality relation it collapses to three:

$$\vec{A} \cdot \underline{B} = A_1 B_1 + A_2 B_2 + A_3 B_3 \ . \tag{6.4.3}$$

Equation (6.4.3) is remarkable because it tells us that the elementary formula for calculating a dot product, Eq. (1.2.2), holds even in the most arbitrary curvilinear coordinate system (of course the two vectors must be an arrow and a stack, and each must be expressed, without confusion, in terms of its correct basis). This is the first example of what we shall later call the "Grand Algebraization Rule," and to which the next chapter will be devoted.

6.5 How Coordinate Systems Differ

Before continuing with the mathematical development, it will be useful to pause and ask ourselves how familiar systems of coordinates – such as Cartesian, cylindrical, or spherical – fit into our formal description. Thus, for example, we know that a Cartesian system is one in which the surfaces of constant q_1, q_2, and q_3 each comprise a family of parallel planes, with the families perpendicular to each other and spaced by unit distance. From the definitions of Sec. 6.1, we would then describe the arrow basis as consisting of three arrows of unit magnitude perpendicular to each other, and the stack basis as three stacks of unit magnitude perpendicular to each other; what's more, these bases do not vary with position. The volume Δ of the unit cell is, of course, unity.

Yet it is also immediately clear that such a description is, topologically speaking, *absolutely illegal,* since every one of the properties just stated requires the use of a calibrated ruler and/or protractor for verification. Conversely, the only topologically legal way of describing vectors numerically – which is to specify their components in terms of the corresponding basis – turns out in this case to be completely empty. For example, the components of the $\vec{e}_1$ basis vector in the Cartesian system are (1,0,0); but, since components are, by definition, the coefficients of a vector when it is expanded in its basis, exactly the same result will obtain for the components of $\vec{e}_1$ *in any other coordinate system.* It is, in fact, impossible, by local topological concepts alone, to distinguish one system from another while remaining entirely within it. Indeed, it could not be otherwise, since a piece of a Cartesian system could, after all, be deformed into, say, a piece of a spherical system, and a topological description is, by definition, one which is independent of deformations of space.

The situation is different, of course, if one is willing to *straddle* two systems, in that it is perfectly possible to define the *relation between a pair of coordinate systems* by specifying components of the basis vectors of one with respect to the other. If, for example, we wanted to relate the Cartesian system to the spherical one, we might begin with the equations

$$x = r \sin \theta \cos \phi \qquad (6.5.1)$$
$$y = r \sin \theta \sin \phi \qquad (6.5.2)$$
$$z = r \cos \theta \; ; \qquad (6.5.3)$$

there then exist systematic procedures to develop all the necessary relations among the bases of the two systems, procedures to which we shall return. Nonetheless, it is crucial to realize that Eqs. (6.5.1-3) do not *define* either a Cartesian or a spherical system, but only the relation between the two. For all we know, the set $\{r, \theta, \phi\}$ could be referring to Cartesian coordinates, in which case $\{x, y, z\}$ would be very, very weird and unfamiliar, yet perfectly well defined.

6.6 But Off the Record, How Does It Look?

In spite of the caution just expressed, whose importance cannot be overstated, it is still helpful to harness our intuitive knowledge of some of the familiar systems to describe them in terms of the concepts we have been developing. What this means, of course, is that in our mind's eye rulers and protractors *do* exist, as shown by the fact that the first paragraph of Sec. 6.5 does carry a clear visual content, a content that will itself become formalized when we introduce the metric properties of space in Chapter 8. Accordingly, we digress briefly at this point to summarize our intuitive metric knowledge for later reference.

In particular, consider the system which we know as "cylindrical," whose three coordinates are labeled $\{\rho, \phi, \zeta\}$; the corresponding families of surfaces are then:

● For ρ: coaxial, equally spaced cylinders;
● For ϕ: planes which contain the axis, and which are equally spaced in angle;

● For ζ: planes perpendicular to the axis, equally spaced.

Figure 6.4 shows a pair of surfaces from each set and a typical "unit cell" formed from their intersections; using the definitions (6.3.3-4), we can then visualize the corresponding basis vectors. The simplest of those are $\vec{e}_\zeta$ and $\underline{\varepsilon}_\zeta$: since the ζ-surfaces are plane and uniformly spaced, those two are constant both in direction and magnitude. By contrast, the ρ-surfaces are uniform in spacing but not constant in direction; accordingly, $\vec{e}_\rho$

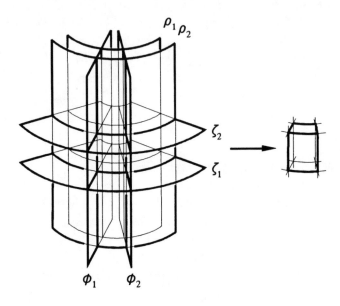

Fig. 6.4: The cylindrical unit cell

and $\underline{\varepsilon}_\rho$ are constant in magnitude but not in direction. Finally, the two φ-vectors are not constant either in direction or in magnitude, with $\vec{e}_\varphi$ varying in proportion to the distance from the polar axis, and $\underline{\varepsilon}_\varphi$ in inverse proportion. (The reader who does not find these statements obvious should reread Sec. 6.3.)

The dependence of basis vectors on position, both with regard to magnitude and direction, is much more the rule than the exception (though it should be noted that this property immediately distinguishes them from what are called "unit vectors" in more elementary treatments). What is, however, special about our particular coordinate system is that, in all cases, the arrow vectors are directed perpendicularly to the sheets of their corresponding stack vectors; or, if we were to specify stack direction by a line normal to its sheets, we could say that the arrow vectors are parallel to the stack vectors. This property characterizes a very special category of coordinate systems known as *orthogonal* ones (to which the cylindrical system does belong).

To repeat: the observations of this section were intuitive, based on our visual ideas of space, and not capable of being formalized in terms of the topological concepts that we have so far admitted into our discussion. They will, however, become important in Chapter 8, when the metric properties of space are more systematically introduced.

PROBLEMS

6.1   Imagine that we change our coordinate system $\{ q_1 , q_2 , q_3 \}$ to a new one in which q_3 varies much more quickly in space than it used to. Will that make the basis vector $\vec{e}_3$ larger or smaller than it was?

6.2   In the change described in the previous problem, will the basis vector $\underline{\varepsilon}_3$ get larger or smaller?

6.3   Continuing from the previous two problems, will Δ get larger or smaller?

6.4   Still continuing, will A_3 in Eq. (6.4.3) increase, decrease, or remain the same? What about B_3? What about the product $A_3 B_3$?

6.5   Show that components of covariant and contravariant vectors $\underline{M}$ and $\vec{N}$ can be computed by the formulas

$$M_i = \vec{e}_i \cdot \underline{M} ,$$
$$N_i = \underline{\varepsilon}_i \cdot \vec{N} .$$

6.6   Are there analogous formulas for calculating components of scalar densities and scalar capacities?

6.7   In the cylindrical coordinate system of Sec. 6.6, how does Δ vary as a function of $\{\rho, \phi, \zeta\}$?

6.8   Repeat our intuitive discussion of the cylindrical coordinate system for the spherical coordinate system $\{r, \theta, \phi\}$.

6.9   Following the discussion of Sec. 6.5, consider the coordinate transformation

$$\xi = x \cos y$$
$$\eta = x \sin y$$
$$\zeta = z.$$

Imagine that $\{x, y, z\}$ are Cartesian. Sketch the contours of constant ξ and of constant η in the z-plane.

7

THE GRAND
ALGEBRAIZATION RULE

7.1 Statement of the Rule

In Sec. 6.4, we noted that the formula for calculating a dot product in terms of components, Eq. (6.4.3), has exactly the same form as the elementary Cartesian recipe, Eq. (1.2.2). In fact, it will turn out that *every single one of the formulas originally listed in Sec. 1.2 preserves its exact form in the most general coordinate system* if we merely replace the Cartesian subscripts $\{x, y, z\}$ with the general subscripts $\{1, 2, 3\}$. We refer to this most remarkable rule, which applies even to the differential operations grad, div, and curl, as the *Grand Algebraization Rule*.

Of course the Grand Algebraization Rule only computes components in terms of components, and will therefore yield nonsense unless we know enough about the quantities in question to be able to associate each component with the correct basis. For example, most standard textbooks on vector analysis state that if the gradient is expressed in, say, spherical coordinates, it is given as

$$\text{grad } \Phi = (\partial\Phi/\partial r)\hat{r} + (1/r)(\partial\Phi/\partial\theta)\hat{\theta} + (1/r\sin\theta)(\partial\Phi/\partial\varphi)\hat{\varphi}; \quad (7.1.1)$$

in this expression the components appear more complicated than the simple partial derivatives to be expected from the Grand Algebraization Rule. The problem is, however, that $\{\hat{r}, \hat{\theta}, \hat{\varphi}\}$ are not our basis vectors. They are "unit vectors," invented by well-meaning people in the belief that in this way they were simplifying the conceptual structure by requiring only one basis instead of four; an advantage that loses much of its appeal when it is realized that it can only be invoked for a tiny subset of possible coordinate systems (the "orthogonal" ones), and that even there its value can be questioned. But if we know (as by now we should) that the gradient is a covariant vector, we shall find that it can indeed be written

$$\text{grad } \Phi = (\partial\Phi/\partial r)\underline{\varepsilon}_r + (\partial\Phi/\partial\theta)\underline{\varepsilon}_\theta + (\partial\Phi/\partial\varphi)\underline{\varepsilon}_\varphi , \quad (7.1.2)$$

as the Grand Algebraization Rule requires.

7.2 The Remaining Bases

When we recall the definition of the cross product of two arrows as giving rise to a thumbtack, and compare it with the pictorial idea of the thumbtack basis as consisting of the three parallelogram faces of the unit cell that come together at a point, it becomes clear that this basis is given by the three possible cross products of the members of the arrow basis, namely

$$\vec{e_2} \times \vec{e_3} , \qquad \vec{e_3} \times \vec{e_1} , \qquad \text{and} \quad \vec{e_1} \times \vec{e_2} . \quad (7.2.1)$$

Analogously, when we recall the definition of the cross product of two stacks as giving rise to a sheaf, and compare it with the pictorial idea of the sheaf basis as consisting of the three sets of parallel edges of the unit cell, it becomes clear that the sheaf basis is given by the cross products of the stack basis, namely

$$\underline{\varepsilon}_2 \times \underline{\varepsilon}_3 \ , \qquad \underline{\varepsilon}_3 \times \underline{\varepsilon}_1 \ , \qquad \text{and} \ \underline{\varepsilon}_1 \times \underline{\varepsilon}_2 \ . \qquad (7.2.2)$$

There exists, however, an alternative way that leads to the same vectors: recalling the meaning of multiplying a stack by a scalar capacity (Sec. 4.7), we can see that the thumbtack $\vec{e}_2 \times \vec{e}_3$

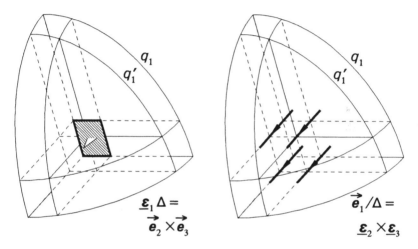

Fig. 7.1: Thumbtack basis and sheaf basis

can also be obtained by multiplying the stack $\underline{\varepsilon}_1$ by Δ. In this way, we arrive at three identities:

$$\underline{\varepsilon}_1 \Delta = \vec{e}_2 \times \vec{e}_3; \quad \underline{\varepsilon}_2 \Delta = \vec{e}_3 \times \vec{e}_1; \quad \underline{\varepsilon}_3 \Delta = \vec{e}_1 \times \vec{e}_2 \ . \qquad (7.2.3)$$

Similarly, recalling what is meant by the product of an arrow by a scalar density, we find three other identities:

$$\vec{e}_1 / \Delta = \underline{\varepsilon}_2 \times \underline{\varepsilon}_3; \quad \vec{e}_2 / \Delta = \underline{\varepsilon}_3 \times \underline{\varepsilon}_1; \quad \vec{e}_3 / \Delta = \underline{\varepsilon}_1 \times \underline{\varepsilon}_2 \ . \qquad (7.2.4)$$

Thus we can specify the thumbtack basis as either $\vec{e}_2 \times \vec{e}_3$ *etc* or $\underline{\varepsilon}_1 \Delta$ *etc,* and the sheaf basis as either $\underline{\varepsilon}_2 \times \underline{\varepsilon}_3$ *etc* or $\vec{e}_1 / \Delta$ *etc.*

These identities provide a way of calculating Δ from either the arrow basis or the stack basis. Consider the first of Eqs. (7.2.3), and take its dot product with $\vec{e}_1$; because of orthonormality, the result is

$$\Delta = \vec{e}_1 \cdot \vec{e}_2 \times \vec{e}_3 \ . \tag{7.2.5}$$

Equivalently, if we take the dot product of the first of Eqs. (7.2.4) with $\underline{\varepsilon}_1$, we get

$$1/\Delta = \underline{\varepsilon}_1 \cdot \underline{\varepsilon}_2 \times \underline{\varepsilon}_3 \ . \tag{7.2.6}$$

Equations (7.2.3-6) also lead to explicit expressions for the stack basis in terms of the arrow basis, and *vice versa:*

$$\underline{\varepsilon}_1 = \vec{e}_2 \times \vec{e}_3 / \vec{e}_1 \cdot \vec{e}_2 \times \vec{e}_3 \qquad etc, \tag{7.2.7}$$

$$\vec{e}_1 = \underline{\varepsilon}_2 \times \underline{\varepsilon}_3 / \underline{\varepsilon}_1 \cdot \underline{\varepsilon}_2 \times \underline{\varepsilon}_3 \qquad etc. \tag{7.2.8}$$

We can now complete our discussion of the computation of dot products in terms of components. Suppose we are given an arrow, a stack, a thumbtack, and a sheaf, each specified in terms of its own basis; that is,

$$\vec{A} = A_1 \vec{e}_1 + A_2 \vec{e}_2 + A_3 \vec{e}_3 \ , \tag{7.2.9}$$

$$\underline{B} = B_1 \underline{\varepsilon}_1 + B_2 \underline{\varepsilon}_2 + B_3 \underline{\varepsilon}_3 \ , \tag{7.2.10}$$

$$\vec{\vec{C}} = C_1 \vec{e}_1 / \Delta + C_2 \vec{e}_2 / \Delta + C_3 \vec{e}_3 / \Delta \ , \tag{7.2.11}$$

$$\underset{\circ}{D} = D_1 \underline{\varepsilon}_1 \Delta + D_2 \underline{\varepsilon}_2 \Delta + D_3 \underline{\varepsilon}_3 \Delta \ . \tag{7.2.12}$$

A simple application of the relationships we have developed among the basis vectors then yields, for the four possible types of dot product, the four formulas

$$\vec{A} \cdot \underline{B} = (A_1 B_1 + A_2 B_2 + A_3 B_3)(1) \ , \tag{7.2.13}$$

$$\vec{A} \cdot \underset{\circ}{D} = (A_1 D_1 + A_2 D_2 + A_3 D_3)(\Delta) , \qquad (7.2.14)$$

$$\vec{C} \cdot \underline{B} = (C_1 B_1 + C_2 B_2 + C_3 B_3)(1/\Delta) , \qquad (7.2.15)$$

$$\vec{C} \cdot \underset{\circ}{D} = (C_1 D_1 + C_2 D_2 + C_3 D_3)(1) . \qquad (7.2.16)$$

Each one of these formulas illustrates the Grand Algebraization Rule: to compute any dot product, *providing it exists,* in terms of the components of the two factors, use the traditional formula (1.2.2) and multiply the result by the correct basis of the type of quantity we are supposed to be getting. Specifically, we multiply by Δ if the answer is to be a scalar capacity, by $1/\Delta$ if it is to be a scalar density, and by 1 (that is, leave it alone) if the product is a pure scalar.

7.3 Cross Products in Terms of Components

In Eqs. (7.2.3-4) we derived the typical cross products of arrow and stack basis vectors. Keeping in mind also the antisymmetry of the cross product, we see that each one of those formulas is a member of a three-by-three table, of which the first is

$$\vec{e}_1 \times \vec{e}_1 = 0 \qquad \vec{e}_1 \times \vec{e}_2 = +\underline{\varepsilon}_3 \Delta \qquad \vec{e}_1 \times \vec{e}_3 = -\underline{\varepsilon}_2 \Delta$$

$$\vec{e}_2 \times \vec{e}_1 = -\underline{\varepsilon}_3 \Delta \qquad \vec{e}_2 \times \vec{e}_2 = 0 \qquad \vec{e}_2 \times \vec{e}_3 = +\underline{\varepsilon}_1 \Delta \qquad (7.3.1)$$

$$\vec{e}_3 \times \vec{e}_1 = +\underline{\varepsilon}_2 \Delta \qquad \vec{e}_3 \times \vec{e}_2 = -\underline{\varepsilon}_1 \Delta \qquad \vec{e}_3 \times \vec{e}_3 = 0 ,$$

and the second

$$\underline{\varepsilon}_1 \times \underline{\varepsilon}_1 = 0 \qquad \underline{\varepsilon}_1 \times \underline{\varepsilon}_2 = +\vec{e}_3/\Delta \qquad \underline{\varepsilon}_1 \times \underline{\varepsilon}_3 = -\vec{e}_2/\Delta$$

$$\underline{\varepsilon}_2 \times \underline{\varepsilon}_1 = -\vec{e}_3/\Delta \qquad \underline{\varepsilon}_2 \times \underline{\varepsilon}_2 = 0 \qquad \underline{\varepsilon}_2 \times \underline{\varepsilon}_3 = +\vec{e}_1/\Delta \qquad (7.3.2)$$

$$\underline{\varepsilon}_3 \times \underline{\varepsilon}_1 = +\vec{e}_2/\Delta \qquad \underline{\varepsilon}_3 \times \underline{\varepsilon}_2 = -\vec{e}_1/\Delta \qquad \underline{\varepsilon}_3 \times \underline{\varepsilon}_3 = 0 .$$

These identities can now be applied to the computation of the cross product in terms of components. For example, let two arrows be expressed as

$$\vec{A} = A_1\vec{e_1} + A_2\vec{e_2} + A_3\vec{e_3} \, , \tag{7.3.3}$$

$$\vec{B} = B_1\vec{e_1} + B_2\vec{e_2} + B_3\vec{e_3} \, . \tag{7.3.4}$$

Their cross product, which from the distributive law would be expected to have nine terms, collapses considerably because of the above tables, yielding the result

$$\vec{A}\times\vec{B} = (A_2B_3 - A_3B_2)\underline{\varepsilon}_1 \, \Delta$$
$$+ (A_3B_1 - A_1B_3)\underline{\varepsilon}_2\Delta + (A_1B_2 - A_2B_1)\underline{\varepsilon}_3\Delta \, , \tag{7.3.5}$$

in agreement with the Grand Algebraization Rule. Similar results are obtained for the other types of cross products.

One further step, which comes out of combining Eqs. (7.3.5) and (7.2.14), produces a useful and elegant expression for the triple scalar product of three arrows in terms of a determinant:

$$\vec{A}\times\vec{B}\cdot\vec{C} = \begin{vmatrix} A_1 & B_1 & C_1 \\ A_2 & B_2 & C_2 \\ A_3 & B_3 & C_3 \end{vmatrix} \Delta \, . \tag{7.3.6}$$

The reader should verify that, if one of the three factors is a sheaf instead of an arrow, the final factor Δ will be missing.

7.4 Hermaphrodite Sense Gender of the Bases

Assuming that the coordinates $\{q_1, q_2, q_3\}$ are ordinary polar scalars (as is usually the case), it follows that the arrow basis vectors $\vec{e_i}$ and the stack basis vectors $\underline{\varepsilon}_i$ are also polar. Since we have defined the thumbtack and sheaf bases as respective cross products of the $\vec{e}_i$ and the $\underline{\varepsilon}_i$, this appears to make the latter two vector sets axial. What, then, are we to do when we wish to write components for, say, a *polar* sheaf or an *axial* stack? Is it impossible to do this without invoking a right-hand rule, thus throwing away invariance under reflection?

Fortunately, such is not the case, because the fact that we are dealing with not two but three vectors provides an alternative method of assigning a sense to these cross products. In particular, we can define $\vec{e}_2 \times \vec{e}_3$ as a thumbtack whose magnitude is the parallelogram subtended by $\vec{e}_2$ and $\vec{e}_3$ (just as before), but whose sense is polar, defined by an arrowhead in the direction of increasing q_1 (that is, in the direction of $\vec{e}_1$ and $\underline{\varepsilon}_1$). An analogous assignment of polar sense can be made for the sheaf $\underline{\varepsilon}_2 \times \underline{\varepsilon}_3$, and axial senses for the arrow and stack bases can be obtained by reversing the process (see Problems 7.3-4). In this way we have, thanks to the presence of three coordinates, succeeded in ascribing to each basis vector both a polar and an axial sense, without ever invoking a right-hand rule. We refer to this property as the *hermaphrodite sense gender* of the bases.

The same considerations apply also to the bases for scalar densities and capacities, since the quantity Δ, the volume of the unit cell, can now be interpreted as either a polar or an axial capacity, depending on the sense gender we assign to the individual vectors in the triple scalar product $\vec{e}_1 \cdot \vec{e}_2 \times \vec{e}_3$.

Any coordinate system can be classified into one of two categories by comparing the hermaphrodite polar/axial relationship of its basis vectors with what would be given by the universal right-hand rule. If the two recipes agree, we say that the system is *right-handed;* otherwise, it is *left-handed.* As shown in Problem 7.6, the handedness of Δ, when viewed as being axial, is the same as that of the system.

7.5 Computation of the Gradient

If the two points 1 and 2 are very close together, Eq. (5.2.2) can be equivalently written as

$$d\Phi = (\text{grad } \Phi) \cdot \vec{dr} , \tag{7.5.1}$$

which can be taken as the analytical definition of the gradient. Since it is a covariant vector, its correct component expansion is

$$\text{grad } \Phi = (\text{grad } \Phi)_1 \underline{\varepsilon}_1 + (\text{grad } \Phi)_2 \underline{\varepsilon}_2 + (\text{grad } \Phi)_3 \underline{\varepsilon}_3 . \tag{7.5.2}$$

As for the differential arrow $\vec{dr}$, it can be written

$$\vec{dr} = dq_1 \vec{e}_1 + dq_2 \vec{e}_2 + dq_3 \vec{e}_3 , \tag{7.5.3}$$

as is perhaps most easily seen from the differential definition, Eq. (6.3.3), but also by considering the geometrical meaning of the arrow basis vectors. Putting the last two equations into the one that precedes them, we obtain

$$d\Phi = (\text{grad } \Phi)_1 dq_1 + (\text{grad } \Phi)_2 dq_2 + (\text{grad } \Phi)_3 dq_3 . \tag{7.5.4}$$

Equation (7.5.4) tells us that $(\text{grad } \Phi)_i$ is *the rate at which Φ changes per unit q_i when the other qs are kept constant;* in other words, it is the partial derivative $\partial \Phi / \partial q_i$. Thus finally the gradient becomes

$$\text{grad } \Phi = (\partial \Phi / \partial q_1) \underline{\varepsilon}_1 + (\partial \Phi / \partial q_2) \underline{\varepsilon}_2 + (\partial \Phi / \partial q_3) \underline{\varepsilon}_3 , \tag{7.5.5}$$

so that its components agree with the Grand Algebraization Rule [cf. Eqs. (1.2.4) and (7.1.2)].

7.6 Computation of the Curl

Our next task — somewhat more complex than the last one — is to calculate the curl of a covariant vector field $\underline{A}$ in terms of its components. In analogy with our treatment of the gradient, we take as our starting point the formula (5.3.4); if the loop around which the line integral is taken is sufficiently small, we can remove the integral sign from the right side and write (at the same time interchanging the two sides of the equation)

$$(\text{curl}\,\underline{A})\cdot d\underline{\Sigma}_{\circ} = \oint \underline{A}\cdot\vec{dr}\,. \tag{7.6.1}$$

The computation becomes especially simple if we choose $d\underline{\Sigma}_{\circ}$ to be one of the three thumbtack basis vectors. Of course this requires justification, since in the equation above $d\underline{\Sigma}_{\circ}$ is supposed to be *infinitesimal;* that is, we are supposed to allow it to approach zero in order to make the equation true. In effect, however, we have already "pre-taken the limit" by assuming that the scale on which the families of q-surfaces are drawn is very fine, so that all the cells become perfect parallelepipeds; we may assume also that the variation of the components of $\underline{A}$ on this

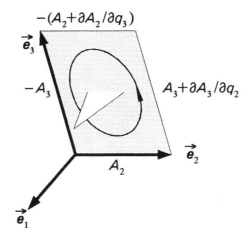

Fig. 7.2: Computation of the curl

scale is very small. Although a more formal treatment of the limiting process is possible, such a "pre-taking of the limit" is in fact absolutely rigorous. After all, the definition of a limit does not really require that the quantity in question become arbitrarily small, but merely that it become small enough so that making it smaller still will not change the result.

Assuming, then, that $d\underset{\circ}{\Sigma}=\underline{\varepsilon}_1\,\Delta$, the left side of (7.6.1) becomes simply $(\text{curl}\,\underline{A})_1$. The right side must still, however, be added up as the sum of the line integrals on the four sides of the parallelogram which bound the thumbtack, as shown in Fig. 7.2. Beginning with the bottom side, we first need the value of $\underline{A}\cdot\vec{e}_2$, which is simply A_2. Since the top side is described in the opposite direction, it would exactly cancel the bottom except insofar as the value of A_2 is not the same there. More specifically, the change of A_2 is exactly its partial derivative with respect to q_3, since the displacement is along $\vec{e}_3$ by unit distance. Thus the combined contribution of the top and bottom is equal to $-\,\partial A_2/\partial q_3$ (negative because the line of higher q_3 is described in the direction opposite to $\vec{e}_2$).

Continuing with the remaining two edges of the thumbtack, we obtain an exactly analogous expression but without the minus sign. When the contributions of all four edges are put together, we finally have

$$(\text{curl}\,\underline{A})_1 \;=\; \partial A_3/\partial q_2 \;-\; \partial A_2/\partial q_3 \;;\qquad\qquad (7.6.2)$$

and when the computation is repeated for the other two components, the result is

$$
\begin{aligned}
\text{curl}\,\underline{A} \;=\;& (\partial A_3/\partial q_2 - \partial A_2/\partial q_3)(\vec{e}_1/\Delta) \\
&+ (\partial A_1/\partial q_3 - \partial A_3/\partial q_1)(\vec{e}_2/\Delta) \\
&+ (\partial A_2/\partial q_1 - \partial A_1/\partial q_2)(\vec{e}_3/\Delta)\,. \qquad (7.6.3)
\end{aligned}
$$

Comparison with Eq. (1.2.6) shows that, once again, the Grand Algebraization Rule triumphs.

7.7 And Finally, the Divergence

Our strategy here is analogous to what we did for the curl: we begin with the integral definition, Eq. (5.4.4), and take one unit cell as the volume in question. Assuming (as we are again allowed to do) that this cell is infinitesimal, we immediately obtain

$$(\text{div}\,\vec{J})\Delta = \iint \vec{J}\cdot d\underset{\circ}{\Sigma} .\tag{7.7.1}$$

The integral on the right has to be summed over the six surfaces of the cell, each of which is one of the basis thumbtacks. Let's

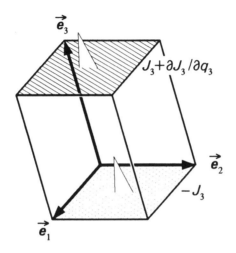

Fig. 7.3: Computation of the divergence

continue to take the equipotentials of the q_3-coordinate to belong to the "floor" and "ceiling" of the cell, as we have consistently been doing. Since the floor is then simply $\underline{\varepsilon}_3\,\Delta$, its contribution to the integral is nothing but J_3 itself, except that the sign must be reversed because, by definition of the divergence, all surface elements $d\underset{\circ}{\Sigma}$ in the integral must point *out of* the volume, whereas our thumbtack points *into* the volume.

Considering now the "ceiling" of the cell, its contribution to the integral will be $+J_3$, without a sign change since $d\underset{\circ}{\Sigma}$ does point out of the cell; but the location at which this J_3 is evaluated is displaced from the other by $\vec{e}_3$ (Fig. 7.3). Accordingly, the net contribution of floor and ceiling will be $\partial J_3/\partial q_3$. When we add the analogous contributions of the other two pairs of faces, and divide both sides of the defining equation by Δ, we find

$$\text{div}\, \vec{\vec{J}} = (\partial J_1/\partial q_1 + \partial J_2/\partial q_2 + \partial J_3/\partial q_3)(1/\Delta) . \qquad (7.7.2)$$

Comparing Eq. (7.7.2) with (1.2.5) may at first produce the impression that the Grand Algebraization Rule is here violated, since the factor $(1/\Delta)$ has no equivalent in the elementary Cartesian formula. We must keep in mind, however, that the divergence of a sheaf field is not a scalar, which would have no basis other than the number 1, but a *scalar density*, which has the basis $1/\Delta$ as in Eq. (7.2.15). And of course the Grand Algebraization Rule does require us, after calculating the component(s) of the desired quantity, to multiply by the corresponding basis − which is precisely what Eq. (7.7.2) correctly does.

PROBLEMS

7.1   Show that an orthonormality relation of the same form as Eq. (6.3.5) exists also between the sheaf and thumbtack basis vectors.

7.2   Derive a determinantal formula analogous to Eq. (7.3.6) for the triple scalar product of three stacks.

7.3   Show that cross products between a thumbtack basis vector and a stack basis vector produce arrow basis vectors.

7.4   Show that cross products between a sheaf basis vector and an arrow basis vector produce stack basis vectors.

7.5   Show that if the quantity Δ is interpreted as being polar, it is necessarily positive.

7.6   Show how the definition of $\underline{\boldsymbol{\varepsilon}}_i$, Eq. (6.3.4), can be viewed as a special case of Eq. (7.5.5).

7.7   Show that in a right-handed system the quantity Δ, when interpreted as being axial, is also right-handed; and conversely for a left-handed system.

7.8 Show, in analogy with Eq. (7.3.5), how the Grand Algebraization Rule applies to the cross product between a thumbtack and a stack.

7.9 To what flavor of quantity does the triple cross product $(\vec{A} \times \vec{B}) \times \underline{C}$ belong? Using components according to the Grand Algebraization Rule, prove the identity

$$(\vec{A} \times \vec{B}) \times \underline{C} = \vec{A} \cdot \underline{C} \, \vec{B} - \vec{B} \cdot \underline{C} \, \vec{A} \, .$$

7.10 Is it possible for a coordinate system to be right-handed in some regions and left-handed in others? Discuss carefully.

8

GOODBYE TO THE
RUBBER UNIVERSE

8.1 The Need for Measurement

Up until now, we have been careful to keep our definitions and formulas topologically invariant. At the same time, we know very well that physics does not take place in a "rubber universe," in which the governing equations have the same form in every possible coordinate system; or equivalently, in which any solution of a physical problem automatically remains a solution regardless of the way in which space is distorted. An elliptical planetary orbit, for example, ceases to be a possible orbit if it is warped into some other shape; clearly, the specification of planetary orbits is not simply topological but also *metric*.

Yet it remains extremely useful, given that some laws of physics cannot be stated in a topologically invariant way while others can, to be able to distinguish the first type from the second, because presumably it is the first that really serve to *define* a particular problem. Occasionally it may even happen that two apparently different physical situations can be formulated so as to differ *only* with regard to topologically invariant specifications; in which case one solution can be obtained from the other by a mere change of coordinates.

8.2 Example: The Electromagnetic Field

The fundamental properties of the electromagnetic field are embodied in the *Maxwell equations:*

$$\text{curl}\,\boldsymbol{E} = -\,(\partial/\partial t)\boldsymbol{B}\,, \tag{8.2.1}$$

$$\text{curl}\,\boldsymbol{H} = (\partial/\partial t)\boldsymbol{D} + \boldsymbol{J}\,, \tag{8.2.2}$$

$$\text{div}\,\boldsymbol{D} = \rho\,, \tag{8.2.3}$$

$$\text{div}\,\boldsymbol{B} = 0\,. \tag{8.2.4}$$

These are further supplemented by two "constitutive relations," which in vacuum take the simple form

$$\boldsymbol{D} = \varepsilon_0\boldsymbol{E}\,, \tag{8.2.5}$$

$$\boldsymbol{B} = \mu_0\boldsymbol{H}\,. \tag{8.2.6}$$

We have used generic boldface letters for all vector quantities so as not to prejudice ourselves before we begin.

Let's now ask what choice of "flavors" would maximize the invariance of the equations. Since a curl can be taken only of a covariant vector field if the relation is to be invariant, (8.2.1) and (8.2.2) suggest that $\boldsymbol{E}$ and $\boldsymbol{H}$ be identified as stack fields $\underline{\boldsymbol{E}}$ and $\underline{\boldsymbol{H}}$. Further, since the curl is itself a contravariant vector density, it would be convenient to view $\boldsymbol{D}$, $\boldsymbol{B}$, and $\boldsymbol{J}$ as sheaf fields $\overrightarrow{\boldsymbol{D}}$, $\overrightarrow{\boldsymbol{B}}$, and $\overrightarrow{\boldsymbol{J}}$. (Remember that, just as in Sec. 2.1, $\partial/\partial t$ is to be considered equivalent to a scalar multiplier.) With these attributions, Eq. (8.2.4) becomes automatically topologically invariant, as does (8.2.3) if we take ρ to be a scalar density. The four Maxwell equations then maintain their validity even under arbitrary coordinate transformations; in other words, given a solution of (8.2.1-4), it remains a solution after any distortion of space.

The situation is radically different, however, with regard to the constitutive relations (8.2.5) and (8.2.6). In particular, if we consider ε_0 and μ_0 to be ordinary scalars, then those two expressions each equate a sheaf field on the left to a stack field on the right, a relationship which is, quite obviously, *topologically illegal* − that is, *metric*. Of course this is not a surprise. For example, in a Cartesian system a possible solution of the electrostatic field in a charge-free region is a (covariant) vector field $\underline{E}$ which has a constant component in one direction (say x) and zero components in the other two directions (y and z). If we try the same thing in cylindrical coordinates, namely a covariant vector $\underline{E}$-field that has a constant component in the radial direction and zero in the two others, it does not work. But there is also a less obvious conclusion that we can draw from these considerations: since the shift from Cartesian to cylindrical coordinates does maintain the validity of Eqs. (8.2.1-4), but violates (8.2.5-6), it follows that the new fields *are* a valid possible solution of the electromagnetic field, *but not in a vacuum*.

8.3 Underlying Cartesian System

As mentioned in Sec. 1.1, the three-dimensional space of our intuition is not only metric but flat, a property that we previously characterized by saying that the sum of angles of an arbitrary triangle is always 180°. More suitably for our present purpose, we may say that such a space possesses an *underlying Cartesian coordinate system* $\{x, y, z\}$, in which the vectors of the arrow basis and of the stack basis are respectively equal:

$$\vec{e}_x = \underline{\varepsilon}_x \, , \qquad \vec{e}_y = \underline{\varepsilon}_y \, , \qquad \vec{e}_z = \underline{\varepsilon}_z \, . \tag{8.3.1}$$

To clarify the meaning of equality between vectors of different flavors, we now formally introduce the two metric notions that we previously encountered on a more intuitive level:

- With regard to direction, vectors whose directional nature is different (line type as against plane type) are considered to have the *same* direction if the line in question is perpendicular to the plane in question;
- With regard to magnitude, vectors can be compared by whatever property defines their magnitude (length of an arrow, sheet density of a stack, area of a thumbtack, or line density of a sheaf), expressed in the *standard units* of our ruler.

So, for example, if the standard ruler is calibrated in centimeters, Eq. (8.3.1) says that the arrow $\vec{e}_x$ is perpendicular to the stack $\underline{\varepsilon}_x$, and the length of $\vec{e}_x$ *in* centimeters is equal to the number of sheets of $\underline{\varepsilon}_x$ *per* centimeter.

Since in this coordinate system, as in any other, we must have

$$\vec{e}_x \cdot \underline{\varepsilon}_x = 1 \,, \tag{8.3.2}$$

it follows that for the underlying Cartesian system the magnitude of $\vec{e}_x$, being the same as the magnitude of $\underline{\varepsilon}_x$, must be 1. Thus the three vectors of the arrow basis, as well as those of the stack basis, must each be equal to unity. Further, since the triple scalar product of the arrow basis is always Δ, and that of the stack basis $1/\Delta$, we conclude that for the Cartesian case $\Delta^2 = 1$, so that Δ is either $\bigcirc 1$ or $\bigcirc 1$ (if viewed as axial), or else $+1$ (if viewed as polar; cf. Problem 7.5).

While we take the definition of a flat space to be one which contains an underlying Cartesian system, it does not mean that this system is unique. For example, if $\{x, y, z\}$ is Cartesian, then a new system defined by

$$x' = x \cos a + y \sin a \,,$$

$$y' = -x \sin a + y \cos a \,, \tag{8.3.3}$$

$$z' = z \,,$$

with a an arbitrary constant angle, is also Cartesian. To prove this assertion, consider x', y', and z' each as a function of $\{x, y, z\}$, and compute their gradients according to Eq. (7.5.5). The result [using the definition (6.3.4)] is that the components of the primed stack basis vectors, given in terms of the unprimed stack basis vectors, are

$$\underline{\varepsilon}_{x'} = (\cos a, \sin a, 0), \tag{8.3.4}$$

$$\underline{\varepsilon}_{y'} = (-\sin a, \cos a, 0), \tag{8.3.5}$$

$$\underline{\varepsilon}_{z'} = (0, 0, 1). \tag{8.3.6}$$

If we now calculate the primed arrow basis vectors by Eq. (6.3.3), we find that their components are the same as those of the primed stack basis; and since in the unprimed system the two bases are identical, the same becomes true in the primed basis. Clearly, the argument can be extended to show that a space which contains an underlying Cartesian system in fact contains an infinity of them.

8.4 Legalization of Illegal Operations; the Laplacian

Because in the underlying Cartesian system the basis vectors of all flavors of vectors are related (being, in fact, equal to each other), operations that were previously "illegal" are no longer so. If, for example, we are called on to find the divergence of a covariant vector, knowing that, topologically speaking, divergences can only be taken of contravariant vector densities, we need only note that in the underlying Cartesian system a covariant vector is equal to a contravariant vector density if they have the same components. In such a case, given the covariant vector

$$\underline{A} = A_x \underline{\varepsilon}_x + A_y \underline{\varepsilon}_y + A_z \underline{\varepsilon}_z, \tag{8.4.1}$$

its divergence is simply taken to be

$$\operatorname{div}\underline{A} = (\partial A_x/\partial x) + (\partial A_y/\partial y) + (\partial A_z/\partial z) \; . \qquad (8.4.2)$$

A much more interesting situation occurs if the divergence of a covariant vector is required in a system which is not Cartesian, but whose relation to a Cartesian system is known. Consider, for example, the coordinate system $\{q_1, q_2, q_3\}$ defined by

$$q_1 = x + y + z \,, \qquad q_2 = y + z, \qquad q_3 = 2z \,, \qquad (8.4.3)$$

or, equivalently, by the inverse set:

$$x = q_1 - q_2 \,, \qquad y = q_2 - \tfrac{1}{2}q_3 \,, \qquad z = \tfrac{1}{2}q_3 \,. \qquad (8.4.4)$$

Using the definition (6.3.4) then yields the following (Cartesian) components for the covariant basis:

$$\underline{\varepsilon}_1 = (1,1,1), \quad \underline{\varepsilon}_2 = (0,1,1), \quad \underline{\varepsilon}_3 = (0,0,2). \qquad (8.4.5)$$

Alternatively, we can obtain the (again Cartesian) components of the contravariant basis by directly applying Eq. (6.3.3) to (8.4.4):

$$\vec{e}_1 = (1,0,0), \quad \vec{e}_2 = (-1,1,0), \quad \vec{e}_3 = (0,-\tfrac{1}{2},\tfrac{1}{2}). \qquad (8.4.6)$$

As for the unit cell volume Δ, we invoke Eq. (7.3.6), knowing that for the Cartesian system this volume is unity:

$$\Delta = \begin{vmatrix} e_{1x} & e_{1y} & e_{1z} \\ e_{2x} & e_{2y} & e_{2z} \\ e_{3x} & e_{3y} & e_{3z} \end{vmatrix} \; ; \qquad (8.4.7)$$

therefore,

$$\Delta = \tfrac{1}{2} \,. \qquad (8.4.8)$$

It is now straightforward to find the Cartesian components of the contravariant density basis vectors:

$$\vec{e_1}/\Delta = (2,0,0), \quad \vec{e_2}/\Delta = (-2,2,0), \quad \vec{e_3}/\Delta = (0,-1,1). \quad (8.4.9)$$

Unlike the Cartesian case, this sheaf basis is not the same as the stack basis; nonetheless, knowing them both allows us to express one in terms of the other, as follows:

$$\underline{\varepsilon}_1 = \tfrac{3}{2}(\vec{e_1}/\Delta) + (\vec{e_2}/\Delta) + (\vec{e_3}/\Delta) \qquad (8.4.10)$$

$$\underline{\varepsilon}_2 = (\vec{e_1}/\Delta) + (\vec{e_2}/\Delta) + (\vec{e_3}/\Delta) \qquad (8.4.11)$$

$$\underline{\varepsilon}_3 = (\vec{e_1}/\Delta) + (\vec{e_2}/\Delta) + 2(\vec{e_3}/\Delta) . \qquad (8.4.12)$$

Taking now any given covariant vector

$$\underline{A} = A_1\underline{\varepsilon}_1 + A_2\underline{\varepsilon}_2 + A_3\underline{\varepsilon}_3 , \qquad (8.4.13)$$

substituting Eqs. (8.4.10-12) into it, and collecting terms, we will find that we have succeeded in writing this stack as an equivalent sheaf — whereupon there is no longer any problem in calculating its divergence.

A particular application of this procedure is in the computation of the *Laplacian* of a scalar function, defined as the divergence of its gradient (clearly a metric operation). Given a scalar function $\Phi(q_1,q_2,q_3)$ of the coordinates defined by Eq. (8.4.3), we obtain its Laplacian by the following sequence of steps:

(a) Take the three partial derivatives of Φ; these are the components of the covariant vector grad Φ.

(b) Convert to a contravariant vector density by using Eqs. (8.4.10-12).

(c) Calculate the divergence of the resulting stack field by Eq. (7.7.2).

The reader may wish to verify that, in this example, the Laplacian will be given by

$$\text{div grad } \Phi = 3(\partial^2 \Phi/\partial q_1{}^2) + 2(\partial^2 \Phi/\partial q_2{}^2) + 4(\partial^2 \Phi/\partial q_3{}^2)$$

$$+4(\partial^2 \Phi/\partial q_2 \partial q_3) + 4(\partial^2 \Phi/\partial q_3 \partial q_1) + 4(\partial^2 \Phi/\partial q_1 \partial q_2) .$$

$$(8.4.14)$$

8.5 The "Del" Operator

We digress for a few moments to discuss the operator ∇, defined as a symbolic vector whose "components" in any coordinate system are simply the three partial derivatives with respect to the three coordinates. The idea is that, using this "vector," the three operations grad Φ, curl $\underline{A}$, and div $\overset{\Rightarrow}{S}$ can be written respectively as $\nabla\Phi$, $\nabla\times\underline{A}$, and $\nabla\cdot\overset{\Rightarrow}{S}$. In the underlying Cartesian system this rather pretty notation is straightforward, since there is no need to ask for the flavor of ∇. But even in a general coordinate system the same expressions lead to correct answers provided that
(a) We think of ∇ as being a *covariant* vector, that is, we write it as $\underline{\nabla}$; and
(b) We apply it only to compute the results of expressions which do, in fact, exist.

As an example of the need for caution, recall that a stack has a valid cross product not only with another stack but also with a thumbtack; also, it has a valid dot product not only with a sheaf but also with an arrow. Yet the corresponding differential operations using $\underline{\nabla}$ — for example, $\underline{\nabla}\cdot\vec{R}$ — are not allowed; that is, the implied combinations of partial derivatives do not yield quantities that can be identified with *any* meaningful geometrical representation, because their nature is metric and not topological.

8.6 Orthogonal Systems

The example that we used in Sec. 8.4 was especially simple in that the linear relation between the qs and the Cartesian coordinates resulted in basis vectors that did not themselves vary with position. As an example of a more complex case, consider the set of "cylindrical coordinates" $\{\rho, \phi, \zeta\}$ defined by

$$\rho = (x^2 + y^2)^{\frac{1}{2}}, \quad \phi = \arctan(y/x), \quad \zeta = z, \tag{8.6.1}$$

or, equivalently, by the inverse transformation

$$x = \rho \cos \phi, \quad y = \rho \sin \phi, \quad z = \zeta. \tag{8.6.2}$$

The reader should note that we do not use the same symbol for the third members of the two sets even though they are equal to each other. The reason has to do with application to partial derivatives, which are meaningless unless one knows not only which quantity is varied in the differentiation but also which others are kept constant; and the common convention is that those quantities are kept constant which belong to the same coordinate system as the one which is varied. So, for example, $\partial/\partial x$ indicates a derivative with respect to x, with y and z kept constant; but if we used the same symbol for z and ζ, we would have no way of knowing whether $\partial/\partial z$ calls for the constancy of x and y or the constancy of ρ and ϕ.

From Eq. (8.6.1) or (8.6.2) we can calculate the various basis vectors of this system, expressed (as in Sec. 8.4) as sets of *Cartesian* components which are, however, now functions of position; as such, they can be specified either in terms of $\{x, y, z\}$ or in terms of $\{\rho, \phi, \zeta\}$, as we prefer. With the latter choice, the stack and sheaf bases are

$$\underline{\varepsilon}_\rho = (\cos \phi, \sin \phi, 0), \tag{8.6.3}$$

$$\underline{\varepsilon}_\phi = (-\sin \phi/\rho, \cos \phi/\rho, 0), \tag{8.6.4}$$

$$\underline{\varepsilon}_\zeta = (0, 0, 1), \tag{8.6.5}$$

and

$$\vec{e}_\rho / \Delta = (\cos\phi/\rho, \sin\phi/\rho, 0), \qquad (8.6.6)$$

$$\vec{e}_\phi / \Delta = (-\sin\phi, \cos\phi, 0), \qquad (8.6.7)$$

$$\vec{e}_\zeta / \Delta = (0, 0, 1/\rho), \qquad (8.6.8)$$

with $\Delta = \rho$.

From these tables it is possible, for example, to express the stack basis in terms of the sheaf basis,

$$\underline{\varepsilon}_\rho = \rho(\vec{e}_\rho / \Delta), \quad \underline{\varepsilon}_\phi = (1/\rho)(\vec{e}_\phi / \Delta), \quad \underline{\varepsilon}_\zeta = \rho(\vec{e}_\zeta / \Delta), \qquad (8.6.9)$$

making it possible, by the method of Sec. 8.4, to write a formula for the Laplacian of a scalar in cylindrical coordinates. The reader should verify that it takes the form

$$\text{div grad } \Phi \equiv \nabla \cdot \nabla \Phi \equiv \nabla^2 \Phi =$$

$$(1/\rho)(\partial/\partial\rho)\rho(\partial\Phi/\partial\rho) + (1/\rho^2)(\partial^2 \Phi/\partial\phi^2) + \partial^2 \Phi/\partial\zeta^2. \qquad (8.6.10)$$

While it is clear that such a method will work whenever a coordinate system is specified in terms of the underlying Cartesian one, the example just given had a rather special feature, in that we were able to write each covariant basis vector in terms of only one contravariant density basis vector rather than all three. In this respect, the cylindrical system represents a special subset of coordinate systems already noted in Sec. 6.6; they are called "orthogonal," because for them the three vectors of the arrow basis are always perpendicular to each other. In this sense, orthogonal systems are like Cartesian ones; but unlike Cartesian ones, their basis vectors need not be of unit length nor of constant direction.

Knowing that for an orthogonal system $\vec{e}_i$ is a multiple of $\underline{\varepsilon}_i$, and that the basic orthonormality, Eq. (6.3.5), applies here as everywhere else, we conclude that for such a system the magnitude of $\vec{e}_i$ is larger than unity by the same factor that the magnitude of $\underline{\varepsilon}_i$ is smaller. This factor is called h_i, and referred to as the *scale factor*. We define also a set of three *unit vectors* $\hat{e}_i$ which point in the same respective directions as the $\vec{e}_i$ (or the $\underline{\varepsilon}_i$)

but have unit size. It follows that

$$\vec{e}_i = h_i \hat{e}_i \, , \tag{8.6.11}$$

$$\underline{\varepsilon}_i = (1/h_i)\hat{e}_i \, , \tag{8.6.12}$$

$$\vec{e}_i = h_i^{\,2} \underline{\varepsilon}_i \, , \tag{8.6.13}$$

all for $i = 1, 2, 3$. Clearly, if a coordinate system is orthogonal, knowledge of the three h_i (each of which is, of course, in general a function of the three coordinates) is sufficient for the type of computation we did in Sec. 8.4; for example, the reader will easily verify that the gradient of a scalar can be computed in terms of unit vectors as

$$\nabla\Phi = (1/h_1)(\partial\Phi/\partial q_1)\hat{e}_1$$
$$+ (1/h_2)(\partial\Phi/\partial q_2)\hat{e}_2 + (1/h_3)(\partial\Phi/\partial q_3)\hat{e}_3. \tag{8.6.14}$$

The formula for the Laplacian is somewhat more complicated, but still straightforward to calculate.

Whether all this is worth the trouble is, to some degree, a matter of taste (see Sec. 6.6). It is true that using three "scale factors" and only one set of unit vectors seems, in a way, much simpler than dealing with four different basis sets; on the other hand, limiting oneself to coordinate systems which happen to be orthogonal is a heavy price to pay, especially as it requires us to abandon the invariance of component equations under coordinate transformations. On the contrary, it can be argued (as this author would) that the geometry-based procedure, though perhaps superficially more complex, is actually a considerable help in maintaining an intuitive grasp of the geometric nature and interaction of the quantities under discussion.

8.7 The Metric

For a general coordinate system (which may or may not be

orthogonal) the concept of scale factor is, of course, meaningless. It turns out, however, that it is possible to formalize the conversion between vector flavors without an explicit specification of how a given system relates to the Cartesian one. Instead, it is sufficient to list the values of the nine dot products that exist among its three arrow basis vectors, that is, the quantities

$$g_{ij} \equiv \vec{e}_i \cdot \vec{e}_j \, . \tag{8.7.1}$$

In fact, only six of them are independent, since, by definition of the dot product, $g_{ij} = g_{ji}$. The quantities (8.7.1) are referred to collectively as the *metric* of the coordinate system, and individually as *components of the metric* (note that in this usage the word "metric" is a noun and not an adjective). In general, all the components g_{ij} are, of course, functions of position, but for a Cartesian system they are given very simply by $g_{ij} = \delta_{ij}$.

The metric can be used to find the distance between two neighboring points in terms of the increments in the coordinates, as follows. One begins with the expression for the incremental displacement $d\vec{r}$, Eq. (7.5.3), and forms its dot product with itself. Then the square of the distance between the points becomes

$$ds^2 = d\vec{r} \cdot d\vec{r} = dq_1 dq_1 \vec{e}_1 \cdot \vec{e}_1 + dq_2 dq_2 \vec{e}_2 \cdot \vec{e}_2 + dq_3 dq_3 \vec{e}_3 \cdot \vec{e}_3$$
$$+ 2dq_2 dq_3 \vec{e}_2 \cdot \vec{e}_3 + 2dq_3 dq_1 \vec{e}_3 \cdot \vec{e}_1 + 2dq_1 dq_2 \vec{e}_1 \cdot \vec{e}_2 \, ; \tag{8.7.2}$$

substitution of Eq. (8.7.1) yields

$$ds^2 = \sum_{i,j} g_{ij} dq_i dq_j \, . \tag{8.7.3}$$

By expanding each $\vec{e}_i$ as a linear combination of the three $\vec{\varepsilon}_k$ and dotting this expression with $\vec{e}_j$, it is easily shown that the coefficients of the linear combination are precisely the quantities g_{ij}, so that

$$\vec{e}_1 = g_{11}\underline{\varepsilon}_1 + g_{12}\underline{\varepsilon}_2 + g_{13}\underline{\varepsilon}_3 \qquad (8.7.4)$$

(plus two analogous equations). This directly provides us with the knowledge required to transform one vector flavor into another, as long as densities or capacities (which require a knowledge of Δ) are not involved; but even this last bit of information can be extracted from the metric, if we recall the following two theorems about determinants:

(a) The value of a determinant is unchanged if one interchanges rows and columns;

(b) The determinant of the product of two square matrices is the product of the determinants of the original matrices.

To apply these theorems, we write Eq. (8.4.7) twice, once as is, and once with rows and columns interchanged, and multiply the two together. According to (b), we then find

$$\Delta^2 = \begin{vmatrix} e_{1x}e_{1x} + e_{1y}e_{1y} + e_{1z}e_{1z} & e_{2x}e_{1x} + e_{2y}e_{1y} + e_{2z}e_{1z} & \cdots \\ e_{1x}e_{2x} + e_{1y}e_{2y} + e_{1z}e_{2z} & e_{2x}e_{2x} + e_{2y}e_{2y} + e_{2z}e_{2z} & \cdots \\ e_{1x}e_{3x} + e_{1y}e_{3y} + e_{1z}e_{3z} & e_{2x}e_{3x} + e_{2y}e_{3y} + e_{2z}e_{3z} & \cdots \end{vmatrix}$$

$$= \begin{vmatrix} \vec{e}_1 \cdot \vec{e}_1 & \vec{e}_2 \cdot \vec{e}_1 & \vec{e}_3 \cdot \vec{e}_1 \\ \vec{e}_1 \cdot \vec{e}_2 & \vec{e}_2 \cdot \vec{e}_2 & \vec{e}_3 \cdot \vec{e}_2 \\ \vec{e}_1 \cdot \vec{e}_3 & \vec{e}_2 \cdot \vec{e}_3 & \vec{e}_3 \cdot \vec{e}_3 \end{vmatrix} . \qquad (8.7.5)$$

Thus *the determinant of the metric is the square of the volume element.* If Δ is regarded as polar, which makes it necessarily positive (Problem 7.5), knowing its square is, of course, equivalent to knowing its value. On the other hand, the question of its handedness and, hence, of the handedness of the coordinate system cannot be resolved from the metric alone.

PROBLEMS

8.1   In hydrodynamics, the equation of continuity of a fluid is

$$\text{div}\,(\rho v) + (\partial/\partial t)\rho = 0,$$

where ρ is the density and v the velocity of the fluid. Can you assign transformation species to these quantities so as to make the equation topologically invariant?

8.2   The equation of motion of the fluid of Problem 8.1 is, for small velocities, given by

$$\rho(\partial/\partial t)v = -\text{grad}\,p,$$

where p is the pressure. Is it possible to extend your assignments so that this equation is topologically invariant?

8.3   "Spherical coordinates" $\{ r,\, \theta,\, \phi \}$ are defined by

$$\rho = (x^2 + y^2 + z^2)^{\frac{1}{2}},\ \ \theta = \arccos\,[z/(x^2 + y^2 + z^2)^{\frac{1}{2}}],\ \ \phi = \arctan\,(y/x);$$

or, equivalently, by the inverse transformation

$$x = \rho \sin\theta \cos\phi,\ \ \ y = \rho \sin\theta \sin\phi,\ \ \ z = \rho \cos\theta.$$

Find the Cartesian components of the four sets of basis vectors, and the quantity Δ, in terms of the spherical coordinates.

8.4   Calculate the Laplacian, in spherical coordinates, of a scalar function $\Phi(r, \theta, \phi)$.

8.5   Show that the spherical coordinate system is orthogonal, and find the three scale factors.

8.6   Given the three components of a contravariant vector $\vec{A}$ in cylindrical coordinates, that is, the coefficients in

$$\vec{A} = A_\rho \vec{e}_\rho + A_\phi \vec{e}_\phi + A_\zeta \vec{e}_\zeta ,$$

find the three components of the contravariant vector $\vec{B}$ defined by $\vec{B} = \text{curl}\,\vec{A}$.

8.7   Repeat for spherical coordinates.

8.8   Suppose we are given the components of the covariant vector $\underline{A}$ when expanded in terms of *unit vectors* of the cylindrical coordinate system. Find the corresponding components of the vector $\overrightarrow{\underline{B}}$ defined by $\overrightarrow{\underline{B}} = \text{curl}\,\underline{A}$. (This is what most books would call "finding the curl in cylindrical coordinates.")

8.9   Find the components of the metric in cylindrical coordinates.

8.10   The same for spherical coordinates.

9

EPILOGUE:
WHERE THIS BOOK LEAVES US

9.1 Some Remaining Problems

The presentation of vectors and vector analysis in this book has, from the beginning, been grounded on the idea that the subject is fundamentally not only geometrical but pictorial; that is, it can be addressed through intuitively familiar notions of a space which is three-dimensional and flat. Yet in spite of its obvious great power, such an approach does have limitations, of which perhaps the most obvious one is the resulting restriction of our menagerie. Specifically, the requirement that every quantity be susceptible to pictorial representation limits us to those that vary proportionally, inversely proportionally, or not at all when space is compressed in some direction; a contravariant vector density, for example, becomes larger in proportion to a sidewise compression, and remains unchanged in a lengthwise compression, and so can be represented as a sheaf. A contravariant vector *capacity,* on the other hand, would have to become smaller in proportion to the *square* of a lengthwise compression. It is difficult to draw a picture which conveniently represents such a quantity; hence the menagerie excludes it.

In this chapter, we shall briefly examine some of the other restrictions which a pictorial treatment imposes and comment on directions that might be taken to rectify them.

9.2 Number of Dimensions

Until now, we have limited ourselves to a space of exactly three dimensions − no more, no less − as being the one with which our intuition is best acquainted. As we know, however, there exist important physical applications in which the dimensionality is different.

Let us look first at the case of only two dimensions, which is already surprisingly unfamiliar. The picture of a stack, for example, now looks very much like the picture of a sheaf; they differ only in that the directional arrowhead of a sheaf points *along* the lines, whereas that of a stack points *across* the lines. A similar situation exists with regard to an arrow and a thumbtack. A sheaf can be transformed into a stack topologically (that is, without ruler or protractor) with the aid of a new kind of right-hand rule; one might say, for example, that the direction of a stack is always *clockwise* from the equivalent sheaf. (Naturally, the arrow-thumbtack pair can be treated similarly.)

The concept of a cross product, too, is radically changed, in that the two-dimensional cross product of a pair of arrows (or stacks) is not a vector at all but a *scalar capacity* (or *density*).

But if the case of two dimensions is one which we perfectly well could have dealt with had we but chosen to, the opposite case − a dimensionality higher than three − presents the more basic problem of the loss of our ability to draw pictures. (Of course "pictures," such as the ones in this book, are actually two-dimensional; but, as we mentioned in Sec. 1.7, the ability to understand three-dimensional space through two-dimensional drawings is built into our intuition − presumably because the retina, a primary tool for perceiving the world, is itself two-dimensional.) As a result, any attempt to investigate a dimensionality higher than three pushes us back onto our mind's ability to extrapolate from the familiar to the unfamiliar, thus taking us to a fundamentally higher level of abstraction.

Consider, for example, the cross product once again. In general, its components are formed as antisymmetric products of pairs of components of the factors, as in Eq. (7.3.5). But in N dimensions, the number of antisymmetric component pairs for a vector is $\frac{1}{2}N(N-1)$, which equals N only for $N = 3$. That is why, in three dimensions, the cross product of two vectors can itself be represented as a vector; but as soon as we go to (say) $N = 4$, the number of its components jumps to six, facing us with an entity we have not previously encountered.

It goes without saying, of course, that a great deal of what we have found out for the case of three dimensions *does* continue to apply to N. Even in those cases, however, we need to be much more careful in our notation.

9.3 Curved Spaces

The methods of Chapter 8, which allowed us to perform complicated operations such as the calculation of Laplacians in general coordinate systems, depended on being able to define any given system by expressing its bases in Cartesian components. But such an approach can only work if the space *contains* an underlying Cartesian coordinate system; in other words, if it is *flat.*

To imagine in our mind a three-dimensional curved space is very difficult indeed. It is much easier to think of a *two*-dimensional curved space, provided it is in turn "embedded" in a three-dimensional space that *is* flat; as, for example, the two-dimensional surface of a sphere. Using the familiar polar coordinates $\{\theta, \varphi\}$, and taking the radius of the sphere to be unity, the square of the distance between two neighboring points becomes

$$ds^2 = d\theta^2 + \sin^2\theta \, d\varphi^2 \ . \qquad\qquad (9.3.1)$$

But to find a new pair of coordinates $\{\theta', \varphi'\}$ which would make the coefficients in Eq. (9.3.1) unity and not bring in any cross terms — which would, in other words, be Cartesian — turns out to be impossible. This means that, for the case of curved spaces, our whole approach to measurement needs to be revised.

9.4 Indefinite Metric

One of the familiar properties of the space of our intuition — so familiar, in fact, that we might easily forget to mention it — is that the square of the distance between two neighboring points, as expressed formally by Eq. (8.7.3), is necessarily positive. Yet the best-known example of a four-dimensional space — the one of Special Relativity, where time is introduced as a fourth dimension — does not satisfy this requirement: as is well known, the invariant square of the space-time interval is there given by the expression

$$ds^2 = dx^2 + dy^2 + dz^2 - c^2 dt^2 \qquad\qquad (9.4.1)$$

(or perhaps its negative). We see that ds^2 takes on one sign for intervals that are "time-like" and the opposite sign for those that are "space-like." Two points ("events") that define the emission and detection of the same light pulse have zero interval between them, even though they can be separated by an arbitrary distance in space (and, of course, also in time).

In this case, the closest one can come to an underlying Cartesian system is one for which, of the four contravariant basis vectors, one produces not $+1$ but -1 when dotted into itself, implying a length that is pure imaginary. Whether such a system is to be described as "Cartesian" at all is, of course, a matter of definition; but it is surely a far cry from the familiar space of our intuition. Hence here, too, the approach that we have used until now requires substantial revision.

It is also worth mentioning that the introduction of time as a fourth dimension means that we can no longer regard time derivatives as equivalent in their transformational properties to multiplicative scalar parameters. The consequences of such a change are most striking in physical laws which are "naturally relativistic," such as the Maxwell equations. In particular, the assignments of three-dimensional vector flavors such as we made in Sec. 8.2 no longer work, but the conclusion that Eqs. (8.2.1-4) are topologically invariant − that is, that they can be written so as to maintain their form under general coordinate transformations − turns out to remain valid.

9.5 The Nature of Tensor Analysis

The fact is that, as Sec. 8.7 already hinted, an alternative method for dealing with the problems of measurement does exist, one that does not require transforming to a Cartesian system but receives its information directly from the metric components g_{ij}; it does, however, require the development of a different, less pictorial formalism. Such a formalism is provided by a field of mathematics where a dazzlingly elegant notation is associated with an approach in which basis vectors are hardly, if ever, mentioned. It is called *tensor analysis.*

One should not make the mistake of thinking that tensor analysis is merely an extension of vector analysis because, even though a closely related mathematical universe is addressed, the fundamental approach of the two formalisms is radically different. As we have repeatedly seen, vector analysis concentrates on properties of geometrical quantities that can be specified without regard to coordinate systems or components; by contrast, the ethos of tensor analysis is to talk *primarily* about components, so that the "underlying geometrical object itself" tends to move into the background. The fact that a relation among tensor quantities is independent of space distortions is then customarily shown, not by avoiding the use of components, but by demonstrating that the corresponding relation among components *maintains its form under general coordinate transformations.* As a result, tensor analysis, while still profoundly geometrical, is not particularly pictorial at all.

In this new formalism, the rôle of arrows, sheaves, and so on is taken by a large genus of quantities called *tensors.* They include, for example, not merely covariant vector *capacities* (our old thumbtacks) but also covariant vector *densities* (for which no simple picture exists). In addition, whereas vector components are numbered by a single index which ranges over the dimensions of the space, a tensor component may carry any number of indices having that same range. Each index can be *covariant* (written as a subscript) or *contravariant* (written as a superscript). One should mention also that there are a number of players on the stage of tensor analysis which are not, in fact, tensors, although they carry similar subscripts and superscripts.

9.6 Conclusion

When all is said and done, the foregoing enumeration of defects of the pictorial approach is no more than an indication that learning has no end. After all, the goal that we set at the beginning of this book was to pursue an understanding of vectors by harnessing our intuitive familiarity with the "normal," three-dimensional Euclidean space in which we conceive of ourselves as living, and this we have done. In the process, the overall nature of vectors and vector operations has been clarified, and many loose ends inherent in more traditional approaches have fallen into place. In that sense, our original goal has been attained.

As for the rest — there is, of course, always more, and we wish the reader all the best in pursuing it!

INDEX